# EATING OFF THE GRID

## storing and cooking foods without electricity

*Denise Hansen, MS, RD*

**EATING OFF THE GRID:**
**storing and cooking foods without electricity**

By Denise Hansen, MS, RD
Cover design by Brenda Brooks


Published by: Subito Services
Post Office Box 28955
Spokane, WA 99228-8955
Email: eatingoffthegrid@usa.net
Http://browser.to/eatingoffthegrid

Additional copies of this book can be obtained by sending checks or money orders, $22 per book (Washington state residents, $23), includes postage and handling, with shipping information to the above publisher. Volume discounts available. Visit the website or send e-mail for additional information.

ISBN 0-9671394-0-6 (pbk.)

Library of Congress Catalog Card Number: 99-93816

Printed in the United States of America.
First printing April 1999.

# ACKNOWLEDGEMENTS

A special thanks to my husband, Bob, for being so supportive and for dealing with all of the technical aspects of this project; to him and our children, for putting up with my cooking experiments; to Bob's mother, for generous use of her computer and taking care of us while we worked; to Dr. Kay Franz, for her mentoring, encouragement and editing help; to Larry Shook and Judy Laddon for sharing their publishing experience and cheering us on; and to our dear friends who volunteered as guinea pigs in the recipe-testing process, the Fletcher's and the Wieber's.

Thanks to William Connor, MD, and Sonja Connor, MS, RD for some of the recipes from their book, *The New American Diet*, published 1986 by Simon and Schuster, which were adapted for this work. Appreciation also goes to the USA Dry Pea and Lentil Council for kind use of the split pea and lentil recipes, which were also adapted. For more information and recipes, the reader may find the Council on the web at www.pea-lentil.com. Or you may send a self-addressed, stamped envelope to:

USA Dry Pea and Lentil Council
5071 Highway 8
Moscow, ID 83843

# CONTENTS

# PREFACE

My viewpoint for the development of this cookbook is summed up in a word picture. Contrast the lifestyles of two distinct groups: condemned prisoners and the nobility of Victorian England. They appear at first to have little in common, yet their diets are actually quite similar. It takes only a change of attitude and a little effort to turn the bread and water rations of condemned prisoners into the elegant toast and tea of ladies and gentlemen of ease. While the prospect of cooking only foods from the pantry over a propane stove can invoke feelings of dread and boredom, it is my hope that you will find the challenge exciting as well as tasty.

These recipes have been collected and developed over a period of about twenty years, during which time my family has been very patient and forgiving, and mostly well-nourished. It began during my graduate work in nutrition at Brigham Young University, when I was asked to design and feed a "food storage" diet to a dozen college students to determine its acceptability and health effects. Recipe development continued through many years of using bulk foods and living off the land to one degree or another. It has been a wonderful time of discovery and re-discovery.

## Who Needs This Cookbook?

This book is primarily intended for those who, by choice or necessity, have no electricity for storing or cooking foods. This will eliminate many appliances our society has come to depend on: refrigerators, freezers, standard stoves and ovens, toasters, blenders, mixers and on and on. However, the recipes may also be used by anyone who is still using such appliances. If you like to eat vegetarian, like to cook from scratch, like to buy and use bulk foods, or like to save money, this book is also for you.

## What Kinds of Foods Are Used?

Foods that could not be used in these recipes include fresh or frozen meats; fresh milk and milk products, including butter, margarine and cheeses; and fresh eggs. It is assumed that you may want to use basic foods in the form available from the farmer, so whole grains and beans are used whenever possible. Fruits and vegetables are intended to be home-grown or obtained locally and used in season; or fruits and vegetables may be canned, dried or root cellared for out-of-season use as well. Powdered milk and powdered eggs are used where necessary. Shortening, oil and peanut butter provide vegetable sources of fat.

## How is this Cookbook Different?

My background in nutrition and food science, coupled with my time in the kitchen "trenches," has uniquely prepared me to see food storage and preparation from the real life side as well as the scientific side. You will find many places in the following chapters where you are not only told to do something, but you are told *why* you need to do it. This can give you the understanding you need to be successful cooking from scratch.

These recipes really work. They have been developed and tested by a professional, but they have also been tested by real families in real homes. Every effort has been made to include all details to ensure good results.

Ingredient and utensil lists are provided to cover all 270 recipes. There is also a sample 4-week meal plan with an accompanying food list, detailing every food and amount required for a family of four using the meal plan. Menus provide 2000-2500 calories per day and are well balanced. They can provide a starting point for someone new to basic foods.

Not one ingredient requiring refrigeration is needed. Instead, substitutions are provided that can be stored for long periods of time, and recipes are modified to use these new ingredients. This is truly a food storage cookbook.

Lastly, the recipe and book layouts have been designed with you in mind. We understand that cooking from scratch takes more time and more instructions, so we have tried to make it as comfortable as possible. Nutrition analysis and exchanges are included for diabetics and those with special dietary needs.

## Headed Off the Grid

Ready for adventures in cooking? There is real satisfaction in knowing where your food comes from. The pride we feel in our first loaf of bread that is tall enough to make a sandwich cannot be adequately described. May you enjoy the journey as well as the destination!

# COOKING PRIMER

## Alternatives for Cooking Appliances

If no standard electric stove or oven is available, it is assumed that some other fuel and cooking appliance will be used for these recipes. Possibilities include a wood cookstove/oven, propane stove or burner, kerosene stove, white gas stove, wood heating stove with a top flat surface, an open campfire or even a solar stove/oven. Choices for cookstoves are many, but baking choices are less plentiful. There are small, collapsible campstove ovens available at stores that sell camping equipment and larger, sturdier models from a few outlets around the country. These portable ovens are simply metal boxes placed on a stovetop of any sort. Foods are placed on the rack inside the box and baking is accomplished by indirect heat. Most of them have a thermometer (NOT a thermostat!) to tell the cook the temperature on the interior of the oven. It is then up to the cook to regulate the stovetop heat, which in turn regulates the temperature inside the oven.

Perhaps if no factory-made oven can be found, a local sheetmetal worker could be hired to build a suitable "oven." Not much more is needed than a tight, heat-resistant metal box with a rack inside to keep the food away from the direct heat. A thermometer that could detect the interior temperature while being read outside by the cook would be helpful, but even a regular, hang-on-the-rack, inside-the-oven thermometer would suffice.

Another oven possibility is the backyard propane grill, which many have. If the unit has dual burners and controls, one side burner can be turned on and food can be placed on the opposite side in the grill, thus achieving the necessary indirect heat for baking.

One last, old-fashioned alternative for an oven is the Dutch oven of camping enthusiasts and boy scouts. These Dutch ovens, unlike those used in the kitchen, have short legs on the bottom and a rim around the edge of the flat lid. These details provide room for coals to be placed on top and under the bottom of the Dutch oven to supply the baking heat. Since there are already many good Dutch oven cookbooks available, Dutch oven cooking will not be covered in this book. You are encouraged to check libraries and bookstores for such information.

Bear in mind that cooking with any of these alternative ovens or stovetops will not be as simple as the usual kitchen appliance. You, not the thermostat, must keep the heat source regulated to maintain a fairly constant oven or stovetop temperature. At first, this may be tricky, but, with practice, you will find yourself enjoying the "art" of cooking, of getting to know your appliance like an old friend, its habits and subtle nuances.

## Alternatives for Cooling Appliances

Learning to do without cooling appliances, refrigerators and freezers may, in fact, require more changes in our cooking habits than doing without cooking or other appliances. If we are used to cooking in large batches to freeze extra meals, or refrigerating leftovers for later use, or pulling a package of meat out of the freezer any time we want to, we may need to change our eating habits and down-size our recipes. One "large batch" habit that will still be possible is to cook a large batch of a simple staple food, such as rice or beans, and then use it in different recipes in two consecutive meals.

Salad dressings present a special problem because these are usually prepared in large enough amounts that they may last a week or two in a refrigerator before being used up. Recipes with perishable ingredients have intentionally been downsized in this book to leave as little leftovers as possible. Pay close attention to recipes to know which are perishable and which are not.

Our forefathers had some ingenious methods for keeping foods cool. The easiest, of course, is to use the outside temperatures whenever possible. Winter months in many locations will allow you to keep foods cool or even frozen outside. Care must be taken to insure that no animals can get into whatever container the food is in. Food will stay colder on the north side of a build or shelter (in the Northern Hemisphere), where it would be protected from winter sun and receive the cooling blast of any northern winds.

Earth or water can also cool foods. Temperatures about six feet underground remain at a fairly constant 55 degrees. Although this temperature is not safe for long storage, it is much better than the air temperature of a summer day and would do for between meal storage. All that is needed is a deep hole, a safe container in which to keep the food, and a rope or other apparatus to raise and lower the container in the hole, plus some good insulation over the top of the hole. Don't forget some firm protection over the top to keep someone from falling in!

Water in an open well, spring or even lake could keep food cooler than air temperatures. Water from deep within the earth, as a well or spring, is especially cool. This method requires a waterproof food container and a way to secure the food from sinking or floating away. Evaporating water can also keep things very cool, as a good breeze after swimming can tell you. A small structure must be built, preferably in a shady spot. It should have a flat roof and floor with open sides, just large enough to hold the amount of food you think you'll need to cool. The entire structure should be draped in burlap or some similar fabric to cover the sides and hang down. Wet the fabric to make it damp. On top of the structure, place upside down a large, lidded jar, filled with water and with holes punctured in the lid. These holes should permit just enough water to leak out to keep the fabric damp. This moisture will slowly evaporate and keep the food inside cooler than air temperature.

Colonial America used the springhouse and icehouse to keep large quantities of food cold. These require the right location and climate, as well as major building projects. The book, *Diary of an Early American Boy*, by Eric Sloane, or one of his other works, gives precise descriptions and drawings of such structures. The *Foxfire* books are another good source of such information.

## Food Safety

A discussion of keeping perishable foods cool would not be complete without some discussion of the basics of food safety. The problem of lack of refrigeration brings with it the inherent increased risk of bacterial growth in foods and subsequent food-borne illness. Understanding the basics of food safety and following appropriate food handling techniques will minimize these risks. The local Cooperative Extension Service has many publications to help with this, so only a few instructions will be provided in this text.

Bacterial growth is greatest in the TEMPERATURE DANGER ZONE, between 40 degrees and 140 degrees Fahrenheit. The general rules for perishable foods are:

1. KEEP HOT FOODS HOT, above 140 degrees.
2. KEEP COLD FOODS COLD, below 40 degrees.
3. IF FOOD MUST BE TAKEN THROUGH THE DANGER ZONE, DO IT AS QUICKLY AS POSSIBLE.

Ideally, one would have a thermometer in the food storage area or container. Without a thermometer, the best that can be done is estimate. 40 degrees and below is typical refrigerator temperature. Since water freezes at 32 degrees, if the temperature in the food container keeps water "ice cold" or even slushy, it could be assumed that the container temperature was less than 40 degrees. 140 degrees is scalding temperature, so safe hot temperatures could not be touched for long without scalding skin. To take the food quickly through the DANGER ZONE, it is important not to leave leftover food just sitting out for long periods of time. It should, instead, be put in the food cooler as soon as possible. If there is a large quantity of food, it should be spread in a wide, flat pan or in several small, shallow pans to increase the surface area and cool it more quickly.

Botulism is a very serious food-borne illness that can be caused by foods of vegetable origin. Since this bacteria can only grow in an environment which contains no oxygen and is not acidic, it can grow in canned foods, particularly home-canned foods if not done properly. Fruits are acidic enough to prevent the growth of botulism, but even today's tomatoes are not acidic enough without the addition of lemon juice in canning. It can even grow in oil, for example, if a clove of garlic (grown in the ground, where botulism bacteria grow) is left in unrefrigerated oil to flavor the oil. Another botulism problem can occur in honey. Botulism spores in honey do no harm to adults (it is the toxin that makes people sick), but the spores are believed to be a cause of Sudden Infant Death. For this reason, honey should never be given to an infant under one year of age. After that time, the more mature intestinal tract can handle the spores just like an adult's does. Again, the local county Cooperative Extension Service can provide more information on this topic.

## How to Use This Cookbook

This section will provide you with lists of all the foods and kitchen tools you will need to prepare these recipes, as well as how to use the recipes to best advantage.

The recipe chapters have been set up not only to simplify meal planning, but also to accommodate the variations in cooking appliances that may or may not be available without electricity. If only a stovetop is available, the chapters on stovetop baking and main dishes may be used. If an oven is also available, then the chapters on baked breads and baked main dishes will offer more variety. The remaining chapters offer a mixed bag of both stovetop and oven requirements, but by far stovetop is most frequent. If a recipe can accommodate either method, it is usually listed in the stovetop category, so check the entire recipe for alternate cooking directions.

Many of the recipes take a good deal of time because they are "from scratch," something that has become a lost art for many of us. Because fuel conservation and time constraints can be very important, the preparation time listed in each recipe takes into consideration not only the actual labor preparation time but also the cooking time. For example, English Muffins has a prep time of 2 hours and 30 minutes. This not only includes the time it takes to mix the dough, but also the rising time and cooking time.

Some of the recipes use pre-cooked ingredients, such as cooked beans or prepared bread dough. The time it takes to prepare these is not included in the prep time, since it is listed as a previously prepared ingredient. It is assumed that the cook would have prepared a large batch ahead of time, planning to use it for more than one recipe. To facilitate this, the index provides lists of those recipes that use various pre-cooked ingredients. Four-alarm Chili, for example, uses cooked pinto beans and kidney beans as well as uncooked lentils. In the index, Four-alarm Chili will be found under "cooked pinto beans" and "cooked kidney beans."

## Ingredient List

All recipes in this book may be prepared from the following list. Seasonings may be changed or omitted according to personal preferences, but it should be noted that the variations in flavor they provide could help prevent the boredom of eating the same foods, a problem for our society spoiled by grocery and fast food stores. Canned and fresh fruits and vegetables may also be modified according to what is available. It is hoped that the reader will grow, or find from local growers, a wide variety of fruits and vegetables for canning, drying, root cellaring and using fresh. These topics are beyond the scope of this book, but much information and help is available through the local Cooperative Extension Service, the library, bookstores and the Internet. No jams or jellies are included in the ingredient list, but these may be made from the fruits and sugar that are listed. Other toppings would be fruit sauce (mashed fruit with a moderate amount of sugar), honey, or cinnamon and sugar.

<u>Grains</u>

- all-purpose flour, optional
- barley
- cornmeal, or whole corn which can be milled into flour
- pasta (spaghetti, lasagna, elbow, rotelli, Ramen, etc.)
- popcorn
- rolled oats
- whole wheat (to be cracked, milled or used whole)

Milk and Animal Protein Foods

- powdered milk (non-instant , low heat type)
- powdered eggs
- canned fish, such as tuna fish (optional)

Sweeteners

- brown sugar
- honey
- molasses
- powdered sugar
- white sugar

Salt

- iodized salt, for eating
- coarse salt, for pretzels only (optional)
- plain salt, optional, for food preservation such as sauerkraut, pickles, etc.

Fats and Oils

- olive oil (recommended, not required)
- pan spray (recommended, not required)
- peanut butter (also supplies some protein)
- vegetable oil
- vegetable shortening

Dried Legumes

- garbanzo beans
- lentils
- navy beans, or other small white beans
- pinto beans
- red beans, or kidney beans
- soybeans
- soy flour (may be milled from beans)
- split peas
- textured vegetable protein (TVP), chunk, cube and granular; may be flavored or plain

Miscellaneous

- baking powder
- baking soda
- bouillon granules: beef, chicken, ham or bacon
- catsup

- chocolate chips
- cocoa
- cornstarch
- Instant Clearjel (cornstarch which thickens without cooking), optional
- mustard
- nuts (optional, if available locally)
- soy sauce
- Tabasco sauce
- vinegar
- Worcestershire sauce
- yeast

Fruits and Vegetables (fresh, canned or dried, as available)

- apples
- applesauce
- berries (such as blueberries, raspberries, strawberries)
- broccoli
- cabbage
- carrots
- chili peppers (jalapeno,chiles)
- corn, canned
- cucumbers
- eggplant
- green beans, canned
- green onions
- onions
- peaches
- peas
- peppers, green and/or red
- plums
- potatoes
- potato flakes,instant (optional)
- pumpkin
- spinach
- Swiss chard, or other greens
- tomatoes
- tomatoes, canned
- tomato sauce
- zucchini

Seasonings (most frequently used listed first and starred)

- *allspice
- *basil
- *chili powder
- *cinnamon
- *cloves
- *cumin
- *dry mustard
- *garlic powder (or fresh)
- *nutmeg
- *oregano
- *parsley
- *pepper, black
- bay leaves
- caraway seed
- cayenne
- celery seed
- chives (or green onions)
- cilantro
- coriander
- dill weed
- dill seed
- fennel seed
- mace
- marjoram
- paprika
- rosemary

- sage
- sesame seed
- thyme
- turmeric

## Suggested Cooking Utensils

The following cooking utensils will be helpful in preparing the recipes in this book. Most of these are fairly common and are probably already in your kitchen. Those items marked with an asterisk are less common, or an item for a special preparation technique.

Baking Utensils

- mixing bowls of various sizes
- sturdy wooden spoons
- sturdy wire whisks,various sizes
- *ball-bearing rotary egg beater
- rubber spatulas or scrapers
- measuring cups for dry ingredient
- measuring cups for liquids
- measuring spoons
- muffin tins
- wire cooling racks
- rolling pin
- cookie sheets, with sides
- loaf pans, 8 ½ x 4 ½ x 2 ½
- oblong pans, 8x11 and 9x13
- square pans, 8x8 and/or 9x9
- 8 or 9-inch layer cake pans
- Bundt pan (optional)
- pie pans, 8 or 9-inch
- *steamed bread molds (1 - 2 qt. tube pans or empty 16 or 20-oz food cans)
- *biscuit and/or doughnut cutter
- *cutters for English muffins (empty tuna cans work well)
- *oven thermometer
- *pastry blender
- *unglazed 8x8 clay tiles (for pita bread only)

Cooking Utensils

- 2 and 3 quart saucepans and lids
- 4 and 6 quart pots and lids
- large, deep skillet, 10 or 12-inch
- *pressure cooker, 6 quart or larger, optional
- vegetable peeler
- colanders, strainers,various sizes
- potato masher
- sieve or food mill, to puree beans, etc.
- *candy/frying thermometer
- pancake turners
- tongs
- sharp knives
- *kitchen scale, to 2 lbs., optional
- *metal rack (like that in pressure cooker) to use in 6-qt pot for steamed breads
- *grain mill (if grains are stored whole, to make flour and cracked grain)
- *tool for shredding vegetables, rotary or by hand
- aluminum foil
- waxed paper

- *cheese cloth (for making cheese)
- heavy griddle
- *cloth-covered board for noodles and piecrust

## Measurement Conversions

| | | | |
|---|---|---|---|
| pinch or dash | = less than 1/8 tsp | 2 cups | = 1 pint |
| 3 teaspoons | = 1 tablespoon | 4 cups | = 1 quart |
| 2 tablespoons | = 1/8 cup | 2 pints | = 1 quart |
| 2 tablespoons | = 1 ounce | 4 quart | = 1 gallon |
| 4 tablespoons | = ¼ cup | 16 liquid oz. | = 1 pint |
| 5 ½ tablespoon | = 1/3 cup | 8 liquid oz. | = 1 cup |
| 8 tablespoons | = ½ cup | 4 pecks | = 1 bushel |
| 16 tablespoons | = 1 cup | 60 pounds | = 1 bushel |
| 1 liquid ounce | = 2 tablespoons | 454 grams | = 1 pound |
| 1 cup | = half pint | 28 grams | = 1 oz |

## Weight-Volume Conversions for Grains and Legumes

| | | | |
|---|---|---|---|
| corn, dry | 1 lb = 2 ¼ cups | white rice | 1 lb = 3 cups |
| cornmeal | 1 lb = 4 cups | white flour | 1 lb = 4 cups |
| lentils & peas | 1 lb = 2 ¼ cups | whl wheat flour | 1 lb = 3 ¾ c. |
| pearl barley | 1 lb.= 2 ¼ cups | whole wheat berries | 1 lb = 2 cups |
| rolled oats | 1 lb = 6 cups | wheat, cracked | 1 lb = 2 ¾ c. |
| soybeans | 1 lb = 2 ¼ cups | other legumes | 1 lb = 2 ½ c. |

## Weight-Volume Conversions for Other Foods

| | | | |
|---|---|---|---|
| white sugar | 1 lb = 2 ¼ cups | potatoes | 1 lb = 3-4 |
| brown sugar | 1 lb = 2 ¾ cups | raisins | 1 lb = 3 cups |
| powdered sugar | 1 lb = 3 ½ cups | spinach | 1 lb = 6 cups leaves |
| honey | 1 lb = 1 1/3 cups | spinach | 1 lb = 1 ¾ c. cooked |
| molasses | 1 c. = 13 ounces | tomatoes | 1 lb = 3-4 |
| shortening | 1 lb = 2 cups | cocoa | 1 lb = 4 cups |
| vegetable oil | 1 lb = 2 cups | chocolate chip | 1 cup= 6 ounces |

## Some Cooked Yields

| | |
|---|---|
| cornmeal | 1 cup = 4 cups cooked |
| pearl barley | 1 cup = 3 ½ cups cooked |

| | |
|---|---|
| rolled oats | 1 cup = 1 ½ cups cooked |
| white rice | 1 cup = 3 cups cooked |
| whole wheat berries | 1 cup = 2 cups cooked |
| whole wheat, cracked | 1 cup = 3 ½ cups cooked |
| lentils | 1 cup = 3 cups cooked |
| other legumes | 1 cup = 2 ½ cups cooked |
| TVP | 1 cup = about 2 cups rehydrated |

## Ingredient Substitutions

It is assumed that anyone in this day and age who is living without electricity is already an expert at "making do or doing without." Please feel free to modify the ingredients in these recipes to suit whatever you have on hand. In most cases, the results will be just fine. Tampering with the main ingredients in baked goods is a little risky, but most other ingredients and recipes may be modified. Especially concerning fruits and vegetables, you may substitute what you have on hand for what is listed in the recipe. Try to match the texture or firmness of the original to keep the results as similar as possible.

Because foods requiring refrigeration or freezing are not used in this cookbook, some substitutes had to be provided. **Nonfat dry milk**, the non-instant type, has been used in place of regular fluid milk. In some recipes it appears in its dry state, with water added separately, and in others in must be pre-mixed and is listed as "reconstituted milk." A powdered soy-based milk is recommended for anyone who cannot or chooses not to drink cow's milk. Check the label to see that your milk replacement has about 7 grams of protein per serving. Many brands are lacking completely in protein, and this may be detrimental to your overall diet unless you know how to replace that protein with additional vegetable sources (such as legumes). In general, if you omit the milk from this diet or use a low protein milk substitute, you would have to almost double the amount of legumes in your diet to replace that protein.

**Soy flour** and water is used as an egg substitute in many of the baked goods. In other recipes, **powdered eggs**, in their reconstituted state, had to be used. Soy flour contains protein, fat and lecithin (an emulsifier) like eggs do, and so soy flour makes a fair replacement in baked goods. In other recipes, where the coagulation of the egg is its most important job (as in custards, for example), powdered eggs must be used.

**Textured vegetable protein (TVP)** is soy protein processed to have the look and texture of meat. It is used in some recipes in place of beef, chicken or pork products. Soaking TVP in hot water containing meat-flavored bouillon helps if the TVP is unflavored.

**Swiss chard stalks**, although not exactly a look-alike, have been substituted for celery because it is so much easier to grow and has such a long growing season. The leaves, a spinach look-alike, are an added bonus.

Following is a list of substitutions for these and many other items.

| Ingredient | Emergency Substitution |
|---|---|
| 1 cup fluid milk | 1 cup water + 3 T. powdered milk |
| 1 cup evaporated milk | 1 cup water + 6 T. powdered milk |
| 1cup sweetened condensed milk | ½ cup warm water + ¾ cup sugar + 1 T. dry milk |
| 1 cup buttermilk, sour milk | 1 c. reconstituted milk + 1 T vinegar |
| ¾ cup non-instant powdered milk | 1 ⅓ cup instant powdered milk |
| 1 egg | 1 ½ T. soy flour + 3-4 T. water |
| 1 egg | 1 T. powdered egg + 1 ½ T. water |
| ¼ cup margarine or butter | 2 T. shortening + 1 T oil + 1 T water |
| 1 oz. baking chocolate | 3 T. cocoa + 1 T. shortening |
| 1 cup all-purpose flour (for baking) | 1 cup or more whole wheat flour |
| 1 T. all-purpose flour (as thickener) | 1 ½ tsp. cornstarch |
| 1 cup honey | 1 ¼ cup sugar + ¼ cup water |
| ¾ cup corn syrup | ½ cup water +1½ cup sugar, heat to dissolve |
| ¼ tsp garlic powder | 1 clove garlic, minced or pressed |
| 1 T. fresh snipped herbs | 1 tsp. dried herbs |
| 1 T. active dry yeast | 1 packet active dry yeast |
| 1 c. shortening / butter in baked goods | ¾ c. thick applesauce or fruit puree + ¼ cup shortening |

## Things to Know about Cooking Basic Foods

### Cooking Legumes

Legumes are the vegetable group that have the capacity to take nitrogen from the soil and "fix" it, or turn it into protein within the plant itself. For this reason, these foods are relatively high in protein and are a staple food for people around the world. Legumes include peanuts, lentils, split peas, white beans, pink beans, red beans, kidney beans, black beans, garbanzo beans, soybeans, black-eyed peas, alfalfa, and the list goes on. Dried legumes may be cooked in a saucepan or in a pressure cooker on top of the stove. They require water or other liquid, oil or other fat, and salt.

The beans absorb water or other liquid during cooking. There must be enough liquid to cover the beans so they will all cook uniformly. A small amount of oil or fat is used in cooking legumes to prevent the foaming that occurs and which may cause boilovers. Oil and salt add flavor to the beans. Some cooks find that adding the salt early in the cooking process seems to keep the beans from softening. Others find that adding the salt early doesn't effect cooking time much, but does add more flavor to the beans. You may make your own decision in this regard.

Acidic ingredients, such as tomatoes, vinegar, catsup or chili sauce DO slow the cooking and softening of the beans. DO NOT ADD these ingredients until after the beans have been softened, as specified in the recipes.

A word about flatulence or, in other words, the gas for which beans are famous. It comes about because we lack the digestive enzymes to completely digest all the different carbohydrates in beans. The result is that these undigested carbohydrates continue to travel down the digestive tract to the colon where, unfortunately, our natural bacteria do have the necessary enzymes. When these bacteria digest the food, they produce gas as a byproduct. The products on the market that are used to lessen the gas problem digest the culprit carbohydates before they get to the colon. No other foolproof method has yet been found, although personal experience has been that gradually the body seems to accustom itself to a high-legume diet and the problem resolves. It may be worthwhile to try eating more of the TVP at first, since this does not seem to cause as much of a problem.

One thing that is not recommended is using baking soda to cook beans. Baking soda is an alkaline, the opposite of an acid, so it tends to soften the beans, just like acid foods tends to harden them. Baking soda has not been proven to lessen gas, but it does destroy many of the B vitamins in beans.

The best cookware for beans is a heavy metal saucepan or pot—cast iron, heavy aluminum (unless you are against it for health reasons), or stainless steel. It should also have a tight-fitting lid to keep in moisture.

A pressure cooker may also be used, to shorten cooking time, since the increased pressure inside the pot cooks the food at a higher temperature. Be sure to follow all safety and usage instructions with your pressure cooker when cooking beans. Never fill it more than about half full, as the foaming liquid could force beans up into the pressure tube and prevent normal pressure release at the end of cooking time.

Large beans should always be soaked before cooking. Lentils, split peas and black-eyed peas are those that do not require soaking. There are two methods for the soak: overnight in cool water; or one hour soak, by bringing beans to a boil for two minutes, removing from heat and letting sit for one hour. Sometimes beans that have been stored for a long time develop "hard shell," meaning that the seed coat is dried to the point that it will not soften and absorb water when soaked or cooked. This can be overcome by the quick method of soaking. The two-minute boil seems to soften that hard shell and allow water to penetrate during the subsequent cooking process.

Specific instructions for cooking legumes of all types are given in the chapter on Basic Recipes.

## Grinding Whole Grains

Storing grains in their whole state allows for longer and better storage, as well as better nutrition from the whole grain. However, some equipment must be available to crack or mill the grain to a more usable form. Dried field corn, particularly, will not cook up soft enough to eat in its whole form. It must be cracked or milled to be edible. Most of us who have stored or grown grain will have purchased an electric (if you have that available) or hand-crank grain mill, and it is best to use the manufacturer's directions when grinding grain. Some general guidelines for use are as follows:

1. Stone mills are best for grains low in oil, such as wheat, triticale, rice, rye, oats and barley. They are generally more efficient and do a better job than metal burr grinders for producing fine flour. The high oil content of soybeans and corn tends to glaze the stones so precautions must be taken if these are milled in a stone mill. The oily grain can either be mixed half (or less) and half with a drier grains for grinding, or a batch of oily grain can be alternated with a batch of drier grain.
2. As explained, metal burr grinders are generally less efficient and take more work to produce a fine flour, but they are usually less expensive and can handle the oily grains without a problem. There are some new versions recently developed that are producing fine flour, so it is important to try your grinder, if at all possible, before purchase. Also, the grain can be run through the mill two or three times, each time on a tighter setting, to produce good flour.
3. If cracked grain is desired, either type of mill should be set to a coarse setting. What is produced is generally a mixture of cracked grain and coarse flour. This can be sifted or sieved to separate the two. The coarse flour can be used as a coating for frying foods, or it can be cooked as "cream of whatever" cereal.

4. If no grain mill is available, a do-it-yourself solution is to pound the grain into meal or flour. Three 30-inch lengths of ordinary ¾-inch pipe may be taped or tied together to form a bundle that can pound the grain. The ends of the pipes should not be capped because the force of the air coming down on the grain would tend to blow it away from the ends of the pipes during pounding. Place about one pound (about 2 cups) of grain in the bottom of a large empty food can (about 4-inch diameter works best.) The grain should be about one inch deep. Place the can on a smooth, firm surface, such as a concrete floor. "Bounce" the bundle of pipes on the grain, going up and down about two inches. One pound of wheat or corn will take about 12 minutes. Sifting will separate the flour from the meal.
5. Another alternative to make the whole grain wheat, corn or rice edible without milling is to parch it by heating it a little at a time in a skillet or in a can while shaking it over a flame, hot coals, or stones. The kernels brown and puff slightly when they are done. The parched grain can be eaten as is or it can more easily be pounded into flour or meal than unparched grain.

## Types of Wheat

There are three major varieties of wheat which vary in their protein content and hardness, thus determiningthe type of baked good for which they are best suited for. The most important of the wheat proteins, as far as baking goes, is the gluten, which becomes elastic as it is worked and gives yeast breads their characteristic texture.

Durum wheat is the hardest and is best for making pasta products. Bread wheats are those that are high in gluten, and ideally should be at least 13% protein. You should be able to get such information from the seller of the grain before purchase. Hard red wheat is one example. The higher the protein, the more gluten and the better yeast bread it will produce. Club wheats, or those best for cakes and pastries, are lower in gluten and produce a more tender baked product. Soft white wheat is one of these varieties. All-purpose flour is, as its name implies, actually for all purposes: it is a mixture of both the high and low protein types of wheat, allowing it to do an "OK" job in either the bread or cake categories. Most recipes in this book have been written for and tested using hard red wheat. The term "whole wheat flour" is used in the recipes. Where needed, it is noted when another variety of wheat or flour is preferable. For more details on baking with whole grains, see the chapters on Baked Breads and Stovetop Breads.

# NUTRITION PRIMER

## Basic Foods for One Year

The following food list has been adapted from *Maintaining Nutritional Adequacy during a Prolonged Food Crisis*, by Franz and Kearny, and published by the US Department of Energy; and from *Essentials of Home Production and Storage*, published by the Church of Jesus Christ of Latter-day Saints. It is the basis for the ingredient list of this cookbook. The amounts below will provide one person approximately 2500 calories per day for an entire year, not including the added fruits and vegetables.

- **300 pounds grains** ( such as wheat, rice, oats, barley, corn, rye)
- **75 pounds non-instant powdered milk** (makes about one quart per day)
- **60 pounds sweeteners** (such as sugar, honey, molasses, brown sugar)
- **10 pounds salt** (iodized, for eating; extra plain salt for food preservation)
- **40 pounds fats and oils** (such as shortening, oil, peanut butter)
- **60 pounds dried legumes** (such as lentils, split peas, pinto beans, etc)
- **365 multi-vitamin/mineral supplements**
- **Garden seeds**

## Total Calories

### Why All the Sugar and Fat?

The total calories are 2500 calories per day to allow for increased activity levels and individual variations in size and age. This may seem high, but it is safer to have plenty than to be short. This is not meant to be a weight loss diet, but one that will maintain adults and allow growth in children and adolescents.

The amounts of sugar, fat and salt may appear high in light of all the bad press these items have received in recent years. In reality, these amounts only seem high because there are virtually no hidden sources in this basic food diet. Most sugar, fat and salt are hidden in the prepared foods of our typical diet: soda pop, breakfast cereals, microwave popcorn, crackers, etc. The sugar amount, about 2 ounces per day, is less than half the amount in the average American diet. The salt is 5 grams per day, which is not far from current diet trends. And the amount of fats works out to about 20% of calories, which is $^1/_3$ less than that recommended by the American Heart Association! In reality, these items are necessary in appropriate amounts in a balanced diet of basic foods. Next to water, salt is the most critical to support life. It also provides palatability to what may be a very bland diet.

## What is bulk?

Besides palatability, sugar and fat do something even more important, they reduce the bulk of the diet. "Bulk" means the volume or amount of food to be eaten. To illustrate this, picture a slice of toast, about 67 calories. Now butter it with 2 teaspoons of butter. You have doubled its calories (134 calories with the butter) without changing the toast's size at all. In other words, you have reduced the "bulk" of that meal. Without the butter, you would have had to eat 2 slices of toast for the same calories. Or picture a bowl of one-half cup oatmeal, again 69 calories. Sprinkle 1 ½ tablespoons of sugar on top, stir it in, and again you have doubled the calories without changing the portion size.

This becomes important when you consider a diet like this that is high in whole grains, legumes, fruits and vegetables, which are all "bulky" foods. One cup of oil contains about 1900 calories. The equivalent calories in cooked oatmeal would take a whopping 14 cups! Most of us, especially children and the elderly, would have a difficult time eating enough to meet their calorie needs on an unrefined diet very low in fats and sugars.

Even healthy young adults cannot make the adjustment easily. As a graduate student in nutrition, I was asked for my research to feed a "stored foods" diet to twelve college students for four weeks and to monitor its effects. This diet was 30% fat for the first two weeks and then 20% fat for the final two weeks. The young men lost an average of 4.3 pounds and the young women, 2.7 pounds in 4 weeks. This was *not* intended to be an experiment in weight loss diets! The people were *provided* with a diet that met their calorie needs, but physically they were unable to eat all of it. Their stomachs were full and they were tired of chewing before their calorie needs were met. Most of them reported gastrointestinal problems, including flatulence and frequent, loose stools. On the other hand, one woman lost no weight and reported no ill effects. She said she was used to eating "health foods" and whole grains prior to the beginning of the study.

The message here is that the transition from a "typical American diet" to this basic foods diet must be made gradually to allow time for the body to adjust. As time goes on, the digestive system can handle the increased fiber and increased bulk, but we must not push the system beyond its limits! If you are ready to begin learning to eat basic foods as a larger part of your diet, it is suggested that you start with no more than one meal a day, and gradually increase to all meals.

## Protein Needs

Changing from a mostly animal protein diet to one of mostly vegetable protein requires some understanding of different protein sources and how our bodies use them.

Proteins are made of a number of different basic building bocks, called amino acids. Some of these we can make ourselves from other building blocks, some we have to get from our diet - these are the essential amino acids. If a protein food contains enough of all the essential amino acids for us to make our own body proteins, it is called a complete protein. All animal proteins (meats, fish, milk, cheese, eggs) are complete proteins, with the exception of gelatin. Our bodies can take apart those building blocks that had been cow, rearrange them, maybe change a few, and turn it into a person! What a marvelous system!

Proteins from vegetable sources, like grains and legumes, are called incomplete proteins because they lack a sufficient amount of one or more of the essential amino acids for us to be able to make a complete "person protein." The marvelous system here is that two or more of these incomplete proteins may be combined in one meal to "complement" each other and provide nearly the same protein quality as an animal protein.

### Complementary Proteins

The only trick is to know which ones to combine. Just pretend you are in a Chinese restaurant. You may choose one (or more) from Column A and one (or more) from Column B. That's the cheap plate. Or, if you're on a freer budget, you may choose one (or more) from Column A or B to go with one from Column C, since animal protein sources are generally more expensive than vegetable sources. Here are some examples:

| Column A<br>Grains | Column B<br>Legumes | Column C<br>Animal Foods |
|---|---|---|
| barley | garbanzo beans | beef |
| buckwheat | white, pink, red, black beans | chicken |
| corn | lentils | cheese |
| oats | peanuts | eggs |
| rice | split peas | fish |
| rye | pinto beans | milk |
| wheat | soybeans | pork |
| spelt, quinoa | tofu (from soybeans) | turkey |
| millet | TVP (from soybeans) | wild game |

Recipes in this work have been written with this in mind. Most main dishes use the "one from column A and one from column B" format. And remember all that powdered milk that was recommended—nearly a quart a day. That would complement either the grains or the legumes to provide you with necessary protein, without any cooking or any bulky fiber!

Protein sources do not have to be eaten in equal amounts. A ratio of 3 or 4 portions of grains to 1 of legumes is sufficient to complement the proteins. This works out to, for example, 2 cups of cooked rice (4 x ½ cup) with ½ cup of cooked beans. The ratio is even better for combining grains or legumes with animal foods—even smaller portions of these are needed.

## Vitamins and Minerals

The amounts of foods in the Basic Foods List (only grains, milk, sweeteners, salt, fats and legumes) will provide a nutritionally adequate diet in nearly all nutrients except vitamins A, C and D. These may be supplied by a multivitamin/mineral supplement and/or fruits and vegetables.

Vitamins A and C are in vegetables that can be grown in a home garden: carrots, pumpkin, spinach, greens, broccoli for vitamin A; potatoes, cabbage, tomatoes and strawberries for vitamin C. Grain and legume sprouts are also a good source of both, when they are allowed to grow to the green leaf stage. Legume sprouts, however, should be cooked or stir-fried for two minutes before eating to inactivate a substance which inhibits protein digestion.

The usual source of dietary vitamin D in the American diet is fortified milk. Unfortunately, the non-instant powdered milk, which keeps much better than the instant, is NOT fortified. Unless a person, particularly a child, can be guaranteed 15 minutes per day of sunshine on at least hands and face for the body to make its own vitamin D, it is probably wise to take a supplement.

# BASIC RECIPES

These recipes will show you how to prepare or cook many of the basic foods you will need in the rest of this cookbook.

This is the place to look if you need to know how to reconstitute a dried food, how to cook beans or rice, how to make refried beans, how to puree beans or even how to make a simple cheese. If you have not cooked from scratch much, it would be worth your time to practice some of these basic recipes before beginning recipes from the rest of the book. It will make it much easier in the long run.

A word about the cheese recipe. Processes involving enzymes and/or molds make usual cheeses. These take a lot of time and skill, so this easy alternative has been provided. The acid in the vinegar clabbers the milk protein to make the curds and whey. The oil in the recipe is added to replace the milk fat missing in powdered milk. The oil softens the curd considerably. It is not required to make the cheese, but without it only the firm, pressed cheese can be made because the curd is so hard.

The whey can be recycled in some of the stir-fry cheese recipes, or it can be used as liquid in baking. Whey contains milk sugar, soluble proteins, vitamins and minerals, so we would not want to waste it.

The pepper cheese recipe is especially interesting. It gives instructions for preserving and aging the cheese without refrigeration for up to one month. Keeping the cheese as cool as possible would still be an advantage.

# How To Reconstitute Powdered Milk

Makes 4 servings. Prep Time: 5 minutes

**¾ c powdered milk, non-instant** **3 ½ c cold water**

1. Pour half the water in a clean 1-2 quart pitcher.
2. Add milk powder and stir well with whisk until all dissolved.
3. Add remaining water and whisk again.
4. Flavor is best if allowed to chill for about an hour. Makes one quart.

**Mix One Cup**: Use 1 c water and 3 tbsp powdered milk.
**Used In Baking**: Powdered milk may be added to dry ingredients and water with the wet. No need to use pre-mixed milk in most recipes.
**Instant Powdered Milk**: To make one quart, use 1 ⅓ c powder instead of ¾ c. Mix same as above.

Per serving (excluding unknown items): 119.0 Calories; 6.4g Fat (48.1% calories from fat); 6.3g Protein; 9.2g Carbohydrate; 23mg Cholesterol; 95mg Sodium. Exchanges: 1 Non-Fat Milk; 1 ½ Fat.

# How to Make Evaporated Milk

Makes 1 serving. Prep Time: 5 minutes

**6 tbsp powdered milk, non-instant** **1 c water**

1. Mix as regular milk. This is double-strength milk, which is the same as evaporated skim milk.
2. Makes 1 c evaporated skim milk.

Per serving (excluding unknown items): 238.0 Calories; 12.8g Fat (48.1% calories from fat); 12.6g Protein; 18.4g Carbohydrate; 47mg Cholesterol; 185mg Sodium. Exchanges: 1 ½ Non-Fat Milk; 2 ½ Fat.

# How to Make Buttermilk

Makes 1 serving. Prep Time: 5 minutes

**1 c milk, reconstituted**
**1 tbsp vinegar**

1. Stir together milk and vinegar. Let stand several minutes before using.
2. Use in place of buttermilk or soured milk in recipes.

Per serving (excluding unknown items): 152.0 Calories; 8.1g Fat (47.5% calories from fat); 8.0g Protein; 12.3g Carbohydrate; 33mg Cholesterol; 120mg Sodium. Exchanges: ½ Non-Fat Milk; 1 ½ Fat.

# How to Make Sweetened Condensed Milk

Makes 4 servings. Prep Time: 5 minutes

**½ c warm water**
**¾ c sugar**
**9 tbsp powdered milk**

1. Pour water into 1-quart bowl or pitcher.
2. Whisk in sugar until dissolved. Stir or whisk in milk until dissolved.
3. Makes about 1 c. This is a very concentrated, sweet milk.

Per serving (excluding unknown items): 234.4 Calories; 4.8g Fat (18.1% calories from fat); 4.7g Protein; 44.4g Carbohydrate; 17mg Cholesterol; 68mg Sodium. Exchanges: ½ Non-Fat Milk; 2 ½ Fruit; 1 Fat; 2 ½ Other Carbohydrates.

# How to Reconstitute Powdered Eggs

Makes 1 serving. Prep Time: 5 minutes

**1 tbsp powdered eggs** **1 ½ tbsp water**

1. Mix together with a whisk; or add powdered egg to dry ingredients and add water with liquid ingredients in baking.
2. See substitution list for using soy flour instead of eggs in baking.

Per serving (excluding unknown items): 0.0 Calories; 0.0g Fat (0.0% calories from fat); 0.0g Protein; 0.0g Carbohydrate; 0mg Cholesterol; 1mg Sodium.
Exchanges: Free.

# How to Rehydrate TVP

Makes 4 servings. Prep Time: 5 minutes

**1 c textured soy protein** **1 c beef broth**

1. Textured soy protein is also called textured VEGETABLE protein, hence the name TVP. TVP may be granular, chunk or cubes. It may come flavored or unflavored. Any will work in this recipe. Ham TVP or soy bacon bits are interchangeable.
2. Plain water, beef broth, chicken broth or ham broth may be used.
3. Pour hot broth or water over TVP and let stand about 5 minutes. Bringing it to a boil may help TVP to absorb all the liquid.
4. This is called "rehydrating" the TVP. Afte rehydrating, use like meat.
5. If TVP is to be used in a soup or stew, it may be added dry and the equivalent amount of liquid added to the stew. The TVP will rehydrate during the cooking process.

Per serving (excluding unknown items): 191.3 Calories; 0.5g Fat (1.9% calories from fat); 41.2g Protein; 13.0g Carbohydrate; 0mg Cholesterol; 331mg Sodium.
Exchanges: 1 Grain(Starch); 5 ½ Lean Meat.

# Basic Cooked Beans

Makes 8 servings. Prep Time: 1 hour and 45 minutes

**2 c dried beans**
**water, for soaking**
**5 c water, for cooking**
**2 tbsp oil**
**1 tsp salt**

1. Soak beans in a pot by covering with water to twice the depth of the beans in the pot. (that is, if the beans are one inch deep in the pot, make the soak water two inches above the beans.) Bring to a boil and boil 2 minutes. Remove from heat and let stand 1 hour. Pour off dirty soak water.
2. Cook beans using water, salt and oil listed above, according to chart below. Cover and simmer in a saucepan or cook in a pressure cooker. Cook at 15 pounds of pressure in pressure cooker.
3. One cup raw beans yields about 2 ½ c cooked, so this recipe will yield about 5 c cooked beans.

Bean Cooking Times

| Bean Type | In a Saucepan | In Pressure Cooker |
|---|---|---|
| lima beans, large or baby | ¾-1 hour | not recommended |
| soybeans | ½-3 hours | 12-20 minutes |
| all other beans* | ½ hours | 5-8 minutes |

*(black, pink, white, red, pinto, kidney, garbanzo, etc)

See page 33 for more information on cooking and using soybeans.

*See text for more specific directions on cooking beans and using a pressure cooker. See also your pressure cooker instruction booklet.*

---

Per serving (excluding unknown items): 204.3 Calories; 4.1g Fat (17.5% calories from fat); 11.6g Protein; 31.5g Carbohydrate; 0mg Cholesterol; 278mg Sodium. Exchanges: 2 Grain(Starch); ½ Lean Meat; ½ Fat.

# Basic Bean Puree

Makes 4 servings. Prep Time: 15 minutes

**1 c cooked beans** **¼ c water**

1. Cook pinto, white or soybeans until very tender.
2. Drain, reserving cooking liquid which may be used instead of water listed above.
3. Puree beans through a sieve, food mill or strainer (Victorio or Squeezo brands, for example) or even through a baby food grinder. A blender or food processor would be electric appliances which could be used instead. If using one of these, add water while beans are being processed.
4. Stir in liquid, adjusting the amount added so that the puree is the consistency of canned pumpkin. Makes about 1 cup puree.

Per serving (excluding unknown items): 64.6 Calories; 0.3g Fat (3.5% calories from fat); 4.0g Protein; 12.0g Carbohydrate; 0mg Cholesterol; 1mg Sodium.
Exchanges: 1 Grain(Starch); ½ Lean Meat.

# Basic Lentil or Split Pea Puree

Makes 4 servings. Prep Time: 1 hour

**1 c lentils, or split peas** **2 ½ c water**

1. Bring ingredients to a boil, reduce heat, cover and simmer until lentils or split peas are very tender, about 45 minutes. The longer they cook, the thicker and softer they become. Stir frequently.
2. Cool slightly, but DO NOT DRAIN.
3. Puree using sieve, food mill, baby food grinder or strainer (Victorio or Squeezo brands, for example). If you have electricity, this may be done in a blender or food processor.
4. Use the cooking liquid already in the lentils. Add extra liquid if needed for proper consistency. It should be the consistency of canned pumpkin.

Per serving (excluding unknown items): 162.2 Calories; 0.5g Fat (2.5% calories from fat); 13.5g Protein; 27.4g Carbohydrate; 0mg Cholesterol; 9mg Sodium.
Exchanges: 2 Grain(Starch); 1 Lean Meat.

# Basic Cooked Peas and Lentils

Makes 8 servings. Prep Time: 45 minutes

**2 c lentils**
**5 c water**
**2 tbsp oil**
**1 tsp salt**

1. Lentils, split peas and black-eyed peas cook quickly and DO NOT REQUIRE SOAKING. Use the above recipe for all three types of legumes.
2. DO NOT USE PRESSURE COOKER. It will overcook these tender beans and could force them up into the vent tube and plug it, not allowing pressure to reduce normally.
3. Rinse beans with water to remove dirt.
4. Combine beans, water, oil and salt in saucepan.
5. Cover and simmer to desired tenderness. Black-eyed peas take 1-1 ½ hours.

Lentils and split peas take 30-45 minutes. Cook least time if they are to be used in a salad. Cook until very tender, most time, if they are to be pureed.

Per serving (excluding unknown items): 192.4 Calories; 3.9g Fat (17.6% calories from fat); 13.5g Protein; 27.4g Carbohydrate; 0mg Cholesterol; 276mg Sodium. Exchanges: 2 Grain(Starch); 1 Lean Meat; ½ Fat.

# A Word about Soybeans

Soybeans are a very unique member of the legume family. They contain less starch and more protein and fat than most legumes. The positive side of this difference is that soybeans are very nutritious and can be used to make so many products: oil, TVP, egg replacer, emulsifier, tofu, tempeh, soy sauce, soy milk, soy cheese, etc. Cooked soybeans may be used as a replacement for other beans in any of the recipes in this book. Soy flour can be added to most baked goods, about one tablespoon per cup of flour to increase protein and nutrition. The negative side, as far as family cooking goes, is that soybeans require much more cooking and processing than other beans (see chart, p.31). Most of the soy products listed above require cottage industry sized effort and equipment. For that reason, they are not included in this cookbook. Anyone who would like to use soybeans more completely is encouraged to search in a library or bookstore or on the Internet.

# Refried Beans

Makes 6 servings. Prep Time: 20 minutes

**4 c pinto beans, cooked**
**¼ c chopped onion**
**½ tsp garlic powder**
**1 tsp salt**
**2 tbsp oil**
**1 tsp cumin**
**1 tsp chili powder**

1. Saute onions in oil until clear.
2. Mash half (or more, depending on preference) the beans and add to the onions. Continue to saute 10 minutes, stirring frequently.
3. Add remaining beans and spices, and cook until warmed. Thin with additional water as needed. Any cooked bean or lentil may be substituted.

Per serving (excluding unknown items): 202.2 Calories; 5.3g Fat (22.9% calories from fat); 9.6g Protein; 30.4g Carbohydrate; 0mg Cholesterol; 363mg Sodium. Exchanges: 2 Grain(Starch); ½ Lean Meat; 1 Fat.

# Quick Refried Beans

Makes 4 servings. Prep Time: 20 minutes

**½ c pinto beans, ground**
**2 ½ c water**
**½ tsp salt**
**¼ tsp garlic powder**
**½ tsp cumin**
**1 tsp chili powder**
**1 tbsp onion flakes**

1. Grind beans as finely as possible. Coarser beans will require more water and cooking time. Use any type bean or lentil. Place in medium saucepan.
2. Add water and stir until smooth. Cook over medium-low heat until beans thicken and begin to bubble, about 10-15 minutes.
3. Add seasonings and stir about 5 minutes. Add more water if necessary.

*This is the way to go if you are short on fuel for cooking! It completely eliminates the soak process and shortens the cooking time by over 75%.*

Per serving (excluding unknown items): 89.5 Calories; 0.5g Fat (4.4% calories from fat); 5.3g Protein; 16.9g Carbohydrate; 0mg Cholesterol; 280mg Sodium. Exchanges: 1 Grain(Starch); ½ Lean Meat.

# Stovetop Rice

Makes 4 servings. Prep Time: 45 minutes

**1 c rice**
**½ tsp salt**
**2 c water**

1. Bring water (or broth) to a boil in a 2-quart heavy saucepan.
2. Stir in rice and salt. Bring back to a boil.
3. Turn off heat. Cover with a tight-fitting lid and let sit about 45 minutes, until liquid is absorbed and rice is tender.
4. Instead, rice may be simmered 15 minutes. Then turn off heat, stir and fluff rice with fork, replace lid and allow to sit and steam for an additional 5-10 minutes.

*To make larger amounts of rice, just keep the proportions the same: 1 part rice to 2 parts water, salt to taste.*

Per serving (excluding unknown items): 168.8 Calories; 0.3g Fat (1.6% calories from fat); 3.3g Protein; 37.0g Carbohydrate; 0mg Cholesterol; 272mg Sodium. Exchanges: 2 ½ Grain(Starch).

# Oven-Steamed Rice

Makes 4 servings. Prep Time: 35 minutes

**2 c boiling water**
**1 c rice, uncooked**
**1 tsp salt**

1. Mix ingredients in an ungreased 1-quart or 7x12 casserole.
2. Cover tightly and bake until liquid is absorbed, 25-30 minutes.

**Seasoned Oven Rice**: Use chicken or beef broth instead of water, and add herbs such as parsley, saffron, onion, dill, garlic, etc.
**Spanish Oven Rice**: Use stovetop recipe. Bake, covered, at 375 degrees for 45 minutes in an ungreased 2-quart casserole.

Per serving (excluding unknown items): 168.8 Calories; 0.3g Fat (1.6% calories from fat); 3.3g Protein; 37.0g Carbohydrate; 0mg Cholesterol; 539mg Sodium. Exchanges: 2 ½ Grain(Starch).

# Homemade Noodles

Makes 4 servings. Prep Time: 3 hours

**2 med egg, reconstituted**
**1 tsp salt**
**4 tbsp milk, reconstituted**
**2 c flour (see note)**

1. Combine all ingredients, adding extra flour or milk if needed to make a stiff dough.
2. Roll very thin on a floured, cloth-covered board; let stand 20 minutes.
3. Cut in strips, or roll up loosely and cut into strips. Spread out or hang to dry 2 hours.
4. Cook in boiling water until tender.

NOTE: Try all different types of flour, and see how you like the results—wheat, white, rye, etc. The cloth-covered board makes this such an easy job. Take the time to sew a cover for your cutting board. You'll be glad you did!

Per serving (excluding unknown items): 137.1 Calories; 1.8g Fat (12.1% calories from fat); 5.0g Protein; 24.4g Carbohydrate; 54mg Cholesterol; 287mg Sodium. Exchanges: 1 ½ Grain(Starch).

# Easy Cheese

Makes 6 servings. Prep Time: 1 hour

**2 qt milk, reconstituted**
**½ tsp salt**
**¼ c vinegar**
**3 tbsp oil**

1. In a deep, heavy 6 or 8-quart pot over medium-high heat, combine milk, oil and salt. Cook, stirring occasionally, until milk comes to a full boil, about 15-20 minutes. Milk at bottom of pan may scorch a little. If this is a problem, try double boiler directions shown below.
2. Stir in vinegar and leave on heat without disturbing while curd separates from the whey ( clear and amber in color.) Boil 2 minutes without stirring and then remove from heat.
3. Line a colander with 3 or 4 layers of cheesecloth, or one layer of any clean, lightweight woven fabric. Piece should be large enough to hang over the sides of the colander. Gently pour curds and whey into colander. Whey may be caught in a container under the colander, to be used in other recipes. Allow cheese to drain until most of the whey is gone, about 5 minutes. Scrape cheese towards center. Continue as follows.
**Drained (Soft) Cheese**: Draw together corners of cheesecloth. Twist to wrap cheese tightly. Turn over in colander. Let rest at room temperature for 20 minutes. Use immediately in recipe, or cheese continues to drain.
**Pressed Cheese**: Fold cheesecloth tightly over cheese to enclose, then invert onto a 10x15 rimmed baking pan (to catch the whey that will be pressed out.) Set a 4 or 5-quart pan filled with water onto wrapped cheese. Let stand about 30 minutes. Lift pan off cheese and unwrap. Makes about 1 ¼ cups cheese.
4. If milk scorches badly, next time try making a large double boiler by using a 4-quart pot (containing the milk) on a rack inside an 8-quart pot filled with water up to the level of the milk in the smaller pot. The boiling water surrounding the milk should bring it to a boil without burning the milk with direct heat on the bottom of the pan.

*This simple cheese comes from India, where it is known as Paneer. It is mild and creamy-textured, similar to our old-fashioned farmer's cheese.*

---

Per serving (excluding unknown items): 261.5 Calories; 17.7g Fat (60.1% calories from fat); 10.7g Protein; 15.8g Carbohydrate; 44mg Cholesterol; 337mg Sodium. Exchanges: 1 Non-Fat Milk; 3 ½ Fat.

# Pepper Cheese

Makes 6 servings. Prep Time: 30 minutes

**1 ¼ c Pressed Easy Cheese**
**½ c cracked black pepper**
**1 ¼ c olive oil**
**½ c vinegar**
**2 tsp basil**

1. Shape cheese into 24 round 1-inch-thick cakes.
2. Roll firmly in cracked pepper.
3. Mix together remaining ingredients.
4. Drop cheese balls into mixture. Cover. Keep at room temperature at least 2 days or up to 1 month. Turn container over weekly.

*This process ages the cheese, so it becomes smoother and more pungent with time.*

Per serving (excluding unknown items): 423.5 Calories; 45.3g Fat (92.7% calories from fat); 1.0g Protein; 7.0g Carbohydrate; 0mg Cholesterol; 4mg Sodium. Exchanges: ½ Grain(Starch); 9 Fat.

# How to Renydrate Dried Fruits and Vegetables

Makes 1 serving. Prep Time: 15 minutes

**1 c dried fruit**
**1 c water**
**1 c dried vegetables**
**1 c water**

1. Home-dried fruits should be dried to a moisture content of about 20%. They may be eaten dried, or they may be rehydrated by covering the fruit with boiling water. Wait ten minutes before serving or using in a recipe.
2. Home-dried vegetables should be dried to a moisture content of about 5%, so they take longer to rehydrate than fruits. Use an equal volume of vegetables and water. Boiling water will quicken the process. It may take from 15 minutes to 3 hours to completely rehydrate the vegetable, depending on its texture and thickness.

NOTE: If fruits or vegetables are commercially dehydrated, follow instructions on package for rehydration.

Per serving (excluding unknown items): 251.8 Calories; 0.5g Fat (1.6% calories from fat); 2.6g Protein; 66.4g Carbohydrate; 0mg Cholesterol; 33mg Sodium. Exchanges: 4 1/2 Fruit.

# BREAKFAST

Breakfasts off the grid do not have to be boring, nor do they have to take a lot of time to prepare. They will definitely be lower in saturated fat and cholesterol than the sausage and eggs variety!

Most of these recipes are stovetop, with a couple requiring baking or steaming (Rice Pudding and Bread Pudding.) Some of the hot cereals can be half-cooked the night before and kept cozy overnight so that they are completely cooked by morning. See overnight method, below. Foods from other chapters, such as Muffins, Whole Wheat CoffeeCake or Peach Foccacia may also be considered for breakfast.

If you think you will miss the convenience of cold cereals, you have four choices for cold breakfasts: Granola, Muesli, Morning Rice and Basic Breakfast Popcorn. Much better nutrition and much less cost will be found in these cereals. Breakfast Popcorn was used by the pioneers as a quick breakfast. Perhaps this was the first puffed cereal! Other grains must be ground or cooked for a long time to be edible, but here is a miracle whole grain, ready to eat in less than five minutes!

Lumpy Dick, Squash GriddleCakes, and Biscuits and Gravy are heirloom recipes from many years ago. Give your family a special heritage day, and have them try some of these hearty breakfasts.

To save on cooking fuel for the cooked cereals, try the overnight method. These are started the same as the regular recipe, but instead of simmering for a long period of time, the cooking process is completed by insulating the pot to retain the heat. Four inches of heat-resistant, loose packing material around all sides of the pot is needed for insulation. A cardboard box or laundry basket with a cotton sleeping bag will do. This will keep the pot hot enough to complete the cooking process. Depending on the quantity being cooked, cooking time may be 2 to 4 hours. This hot box can reduce energy consumption by three to twenty times, as well as freeing you from having to watch the stove and stir.

# Blender Pancakes

Makes 4 servings. Prep Time: 30 minutes

**1 c whole wheat berries**
**1 ½ c milk, reconstituted**
**½ tsp salt**
**2 tbsp sugar**
**⅓ c oil**
**2 tbsp soy flour**
**¼ c water**
**1 tbsp baking powder**
**oil, for frying**

1. Put wheat berries and 1 c milk in blender. Blend on high speed 3 minutes.
2. Add remaining ½ c milk and ¼ c water. Blend 2 more minutes.
3. Add remaining ingredients and blend well
4. Cook on a hot, lightly greased griddle, as in regular pancake recipe. Use similar toppings.

*This is the only recipe that DOES REQUIRE ELECTRICITY in the form of a blender. This is also the only recipe I have found that allows you to turn whole wheat berries into flour without a grinder, so I couldn't pass it up! If you don't have a grinder, but you do have the kilowatts, enjoy the pancakes!*

Per serving (excluding unknown items): 373.9 Calories; 22.2g Fat (51.5% calories from fat); 8.2g Protein; 38.8g Carbohydrate; 12mg Cholesterol; 590mg Sodium. Exchanges: 2 Grain(Starch); ½ Non-Fat Milk; ½ Fruit; 4 Fat; ½ Other Carbohydrates.

# Bread Pudding

Makes 4 servings. Prep Time: 1 hour and 20 minutes

**4 slices bread, dried**
**1/3 c brown sugar**
**½ tsp cinnamon**
**2 ½ c milk, reconstituted**
**1/3 c raisins**
**4 ½ tbsp soy flour**
**¼ c cornstarch, dissolved in**
**¼ c water**
**1 tsp vanilla**
**1/3 c sugar**

1. In a greased 1 ½-quart casserole, stir together bread (cut in cubes), brown sugar, cinnamon and raisins.
2. Scald milk in a small saucepan.
3. Stir in soy flour, dissolved cornstarch, vanilla and sugar using a whisk. Pour carefully over bread.
4. Bake in a 350 degree oven, placing casserole in a 9x9 pan and then filling it one inch deep with hot water.
5. Bake 60-70 minutes, until set. Serve warm or cold.

**Steam Baked**: Cover casserole tightly with foil. Place casserole on rack in a Dutch oven. Fill with hot water to just above rack level. Gently steam on stovetop about 1 ½-2 hours, until set. See chapter on Stovetop Breads for further information on steamed "puddings."

Per serving (excluding unknown items): 390.6 Calories; 7.3g Fat (16.5% calories from fat); 9.5g Protein; 73.5g Carbohydrate; 21mg Cholesterol; 220mg Sodium. Exchanges: 1 ½ Grain(Starch); ½ Non-Fat Milk; 1 ½ Fruit; 1 Fat; 2 ½ Other Carbohydrates.

# Breakfast Popcorn

Makes 4 servings. Prep Time: 5 minutes

**8 c popcorn**
**4 c milk, reconstituted**
**4 tbsp sugar**

1. Allow popcorn to cool. Pour into serving bowls.
2. Pour 1 c milk over each and sprinkle with 1 tbsp sugar.
3. Eat quickly before it gets soggy!

NOTE: Any sweetened popcorn recipe from the Theme and Variations list may be substituted for plain popcorn.

Per serving (excluding unknown items): 308.3 Calories; 14.3g Fat (41.0% calories from fat); 10.0g Protein; 36.4g Carbohydrate; 33mg Cholesterol; 314mg Sodium. Exchanges: 1 Grain(Starch); ½ Non-Fat Milk; 1 Fruit; 2 ½ Fat; 1 Other Carbohydrates.

# Cornmeal Mush

Makes 4 servings. Prep Time: 15 minutes

**¾ c cornmeal**
**¾ c water, cold**
**2 ½ c water, boiling**
**½ tsp salt**

1. Mix cornmeal and cold water in a 2-quart saucepan. Stir in boiling water and salt.
2. Cook, stirring constantly until mixture boils and thickens.
3. Reduce heat. Cover and simmer 10 minutes.
4. Serve with honey, sugar or maple syrup.

Leftovers may be used in Fried Mush recipe.

Per serving (excluding unknown items): 94.7 Calories; 0.4g Fat (4.2% calories from fat); 2.2g Protein; 20.1g Carbohydrate; 0mg Cholesterol; 273mg Sodium. Exchanges: 1 ½ Grain(Starch).

# Cracked Wheat Cereal

Makes 6 servings. Prep Time: 30 minutes

**1 ½ c cracked wheat cereal**
**6 c water**
**1 tsp salt**

Bring all ingredients to a boil in a 3-quart saucepan. Cover and simmer 30-45 minutes, until water is absorbed and wheat is tender.
2. Add honey or sugar, raisins or chopped fruit as desired.
**Overnight Method**: Boil only 15-20 minutes. Keep covered and place in hot box. See Breakfast introduction page for directions. Let sit several hours or overnight, until all water is absorbed.

*Any cracked grain may be cooked using this recipe.*
Leftovers may be used in Fried Mush recipe or in your favorite stuffing recipe.

Per serving (excluding unknown items): 119.7 Calories; 0.5g Fat (3.2% calories from fat); 4.3g Protein; 26.6g Carbohydrate; 0mg Cholesterol; 368mg Sodium. Exchanges: 2 Grain(Starch).

# Cream of Rice Cereal

Makes 4 servings. Prep Time: 15 minutes

**1 c rice, coarsely ground**
**4 c water**
**½ tsp salt, opt**

1. To avoid lumping, combine rice flour with 1 c of the water in a separate bowl, using a wire whisk.
2. Meanwhile, bring remaining 3 c of water and salt to a boil in a 2-qt. pan.
3. Stir in flour mixture quickly and cook until thick, stirring frequently.

*Any other grain, ground to a coarse flour, may be used.*

Per serving (excluding unknown items): 168.8 Calories; 0.3g Fat (1.6% calories from fat); 3.3g Protein; 37.0g Carbohydrate; 0mg Cholesterol; 276mg Sodium. Exchanges: 2 ½ Grain(Starch).

# Fried Mush

Makes 4 servings. Prep Time: 25 minutes

**4 c cooked cereal, leftover**
**flour**
**3 tbsp oil, or shortening**
**maple flavored syrup**

1. Spread leftover cereal in a greased loaf pan, 9x5x3 or smaller. Cover and chill until firm, at least 12 hours.
2. Invert pan to unmold. Cut loaf into half inch slices.
3. Heat oil or shortening in 10-inch skillet. Coat slices with four.
4. Cook over low heat until brown on both sides.
5. Serve hot with honey or maple flavored syrup, if desired.

*Any thick, cooked cereal, such as cracked wheat, oatmeal or cornmeal may be used.*

Per serving (excluding unknown items): 0.0 Calories; 0.0g Fat (0.0% calories from fat); 0.0g Protein; 0.0g Carbohydrate; 0.0mg Cholesterol; 0.0mg Sodium. Exchanges: Free.

# Hashed Browns

Makes 4 servings. Prep Time: 30 minutes

**1 ½ lb potatoes, precooked**
**2 tbsp onion, finely chopped**
**½ tsp salt**
**1/8 tsp pepper**
**2 tbsp shortening**
**2 tbsp oil**

1. Use pre-boiled or baked potatoes. Shred enough for 4 c.
2. Mix together potatoes, onion, salt and pepper.
3. Pack firmly into skillet, leaving about ½-inch space around edge.
4. Cook over low heat until bottom crust is brown, about 10-15 minutes.
5. Cut potato mixture into quarters; turn.
6. Cook until brown, about 12-15 minutes until brown on other side.

Per serving (excluding unknown items): 219.8 Calories; 13.4g Fat (53.5% calories from fat); 2.7g Protein; 23.4g Carbohydrate; 0mg Cholesterol; 274mg Sodium. Exchanges: 1 ½ Grain(Starch); 2 ½ Fat.

# Granola

Makes 16 servings. Prep Time: 45 minutes

**12 c rolled oats**
**1 c whole wheat flour**
**1 c wheat bran or oat bran**
**1 ½ c raisins**
**½ c brown sugar**
**½ c oil**
**½ c honey**
**flavoring (see below)**
**¾ c water**

1. Combine oats, flour and bran in a large mixing bowl. Meanwhile, heat brown sugar, oil, honey, flavoring and water in a 1 quart saucepan, until well-blended.
2. Pour liquid over oat mixture and stir until thoroughly moistened.
3. Spread on 2 large lightly greased baking sheets and bake at 300 degrees for about 30 minutes, until cereal is lightly browned and crisp.
4. Stir in raisins after cereal has cooled. Serve cold with milk. Keeps in a tightly covered container on the shelf for several weeks.

**Flavorings**: Choose from one of the following--2 tsp vanilla, maple or almond flavoring; 1 tbsp cinnamon.

**Peanut Butter Granola**: Add ½ c peanut butter to liquids before heating. Stir well to blend in melted peanut butter.

**Popcorn-Ola**: Substitute 2-4 c of medium ground popcorn (ground after popping) for 2-4 c of rolled oats.

Per serving (excluding unknown items): 417.8 Calories; 10.8g Fat (22.6% calories from fat); 11.2g Protein; 72.3g Carbohydrate; 0mg Cholesterol; 8mg Sodium. Exchanges: 3 Grain(Starch); ½ Fruit; 2 Fat; 1 Other Carbohydrates.

# Honey-Frosted Popcorn

Makes 6 servings. Prep Time: 15 minutes

**4 qt popcorn**
**1 c raisins**
**¾ c honey**
**½ c water**

1. Mix together popped popcorn and raisins. Keep warm in a large, greased pan in a 250 degree oven.
2. In a saucepan, boil honey and water for several minutes. Be careful not to burn.
3. Remove popcorn from oven. While honey is still foamy, pour on top and mix thoroughly.
4. Spread in greased pan, and separate into small pieces if possible.
5. Cool and store in a tightly covered container. Use as any cold cereal.

REMINDER: If you have no honey, remember that you may substitute 1 ¼ c sugar plus ¼ c water.

---

Per serving (excluding unknown items): 348.0 Calories; 8.4g Fat (20.2% calories from fat); 3.6g Protein; 70.8g Carbohydrate; 0mg Cholesterol; 265mg Sodium. Exchanges: 1 Grain(Starch); 1 ½ Fruit; 1 ½ Fat; 2 ½ Other Carbohydrates.

# French Toast

Makes 6 servings. Prep Time: 30 minutes

**12 slices bread**
**2 c water**
**½ c powdered milk**
**4 tbsp egg substitute, dried**
**¼ tsp salt**
**1 tbsp sugar**
**2 tbsp oil, for frying**

1. Use thick-sliced, stale bread. Heat some oil in griddle on medium heat.
2. Combine all ingredients except bread and oil using a wire whisk.
3. Dip bread slices in mixture and cook on griddle until crisp and browned.
4. Add oil as needed to prevent sticking. Serve with favorite topping.

---

Per serving (excluding unknown items): 254.4 Calories; 9.8g Fat (34.7% calories from fat); 9.4g Protein; 31.9g Carbohydrate; 36mg Cholesterol; 435mg Sodium. Exchanges: 1 ½ Grain(Starch); ½ Lean Meat; ½ Non-fat Milk; 2 Fat.

# Lumpy Dick

Makes 4 servings. Prep Time: 25 minutes

**2 c milk, reconstituted**
**1 tbsp shortening**
**¼ tsp salt dash pepper**
**1 c whole wheat flour**
**¼ tsp salt**
**2 tbsp powdered eggs**
**½ c water**

1. Cook milk, salt and pepper over medium heat until milk is scalded.
2. Meanwhile, mix remaining ingredients. Drop by spoonfuls into hot milk.
3. Cover, reduce heat and simmer about 10 minutes. May add sugar and cinnamon to taste.

Per serving (excluding unknown items): 242.2 Calories; 10.3g Fat (37.4% calories from fat); 11.3g Protein; 27.8g Carbohydrate; 124mg Cholesterol; 359mg Sodium. Exchanges: 1 ½ Grain(Starch); ½ Lean Meat; ½ Non-Fat Milk; 1 ½ Fat.

# Morning Rice

Makes 4 servings. Prep Time: 20 minutes

**3 c cooked rice**
**1 c milk, reconstituted**
**¼ c sugar**
**1 each apple, chopped**
**1 tsp cinnamon**
**3 tbsp raisins**

1. Mix all ingredients together in a medium bowl.
2. Let stand about 15 minutes.

*Jam or honey could be used in place of sugar.*

Per serving (excluding unknown items): 311.9 Calories; 2.6g Fat (7.5% calories from fat); 6.1g Protein; 66.2g Carbohydrate; 8mg Cholesterol; 34mg Sodium. Exchanges: 2 ½ Grain(Starch); 1 ½ Fruit; ½ Fat; 1 Other Carbohydrates.

# Muesli

Makes 6 servings. Prep Time: 35 minutes

**2 c rolled oats**
**2 c milk, reconstituted**
**2 large apples, chopped**
**½ c raisins**
**¼ c sugar**
**dash cinnamon**

1. Combine all ingredients, except chopped apples, and let stand at least 30 minutes or overnight in a cool place (under 45 degrees).
2. Add chopped apples before serving.
3. Other fruits could be used, or water or juice could be substituted for milk.

*This Swiss oatmeal is prepared and served cold.*

Per serving (excluding unknown items): 263.8 Calories; 4.7g Fat (15.5% calories from fat); 7.5g Protein; 50.5g Carbohydrate; 11mg Cholesterol; 42mg Sodium. Exchanges: 1 Grain(Starch); 2 Fruit; 1 Fat; ½ Other Carbohydrates.

# Oatmeal

Makes 6 servings. Prep Time: 20 minutes

**3 c rolled oats**
**6 c water**
**1 tsp salt**

1. Combine all ingredients in a large saucepan. Bring to a boil. Stir frequently and cook until thick, about 15 minutes.
2. Add honey or sugar, raisins or fruit as desired.

**Elegant Oatmeal**: Add to the above, before cooking 1 chopped apple, ¼ c raisins, 1 tsp cinnamon and ¼ c sugar.

*Any rolled grain may be cooked using this recipe, but cooking times may vary.*

Leftovers may be used in Fried Mush recipe.

Per serving (excluding unknown items): 155.5 Calories; 2.6g Fat (14.6% calories from fat); 6.5g Protein; 27.1g Carbohydrate; 0mg Cholesterol; 364mg Sodium. Exchanges: 2 Grain(Starch); ½ Fat.

# Oatmeal Pancakes

Makes 4 servings. Prep Time: 60 hours

**1 ½ c rolled oats**
**2 c milk, reconstituted**
**2 each eggs, reconstituted**
**1 c whole wheat flour**
**1 tbsp baking powder**
**2 tbsp brown sugar**
**oil, for frying**
**½ tsp salt**
**2 tbsp melted shortening**

1. Combine oats, milk and eggs. Let stand 30 minutes or chill (under 45 degrees) up to 24 hours to allow time for oats to absorb liquid and thicken slightly. The process may be hastened by heating the mixture on the stove in a saucepan until the oats are partially cooked.
2. Add remaining ingredients and stir just until moistened.
3. Cook on a hot, lightly oiled griddle
4. Serve with fruit, jam, honey or maple syrup.

Per serving (excluding unknown items): 358.2 Calories; 9.1g Fat (22.1% calories from fat); 16.1g Protein; 55.6g Carbohydrate; 124mg Cholesterol; 369mg Sodium. Exchanges: 3 Grain(Starch); ½ Lean Meat; ½ Non-Fat Milk; 1 ½ Fat; ½ Other Carbohydrate

# Pancakes

Makes 4 servings. Prep Time: 30 hours

**2 c whole wheat flour**
**¼ c oil, or melted shortening**
**2 tbsp sugar**
**2 tbsp soy flour**
**¼ c powdered milk**
**2 tbsp baking powder**
**1 ¾ c water**
**oil, for frying**

1. Stir together all dry ingredients in a medium bowl.
2. Add water and oil; stir just until smooth.
3. Cook on a hot, lightly greased griddle. Cook on one side until puffed and dry around edges. Turn and cook on other side until golden brown.
4. Serve with fruit, honey, jam or maple syrup.

**Applesauce Pancakes**: Decrease water to 1 ¼ c. Beat in ½ c applesauce with liquid ingredients. Add ½ tsp cinnamon.

**Berry Pancakes**: Stir in 1 c blueberries or other berries.

**Peach Pancakes**: Beat in ½ tsp cinnamon and 2 medium peaches, peeled and cut up.

---

Per serving (excluding unknown items): 402.7 Calories; 17.4g Fat (37.0% calories from fat); 11.2g Protein; 55.4g Carbohydrate; 8mg Cholesterol; 581mg Sodium. Exchanges: 3 Grain(Starch); ½ Non-Fat Milk; ½ Fruit; 3 ½ Fat; ½ Other Carbohydrates.

# Popcornmeal

Makes 4 servings. Prep Time: 15 minutes

**12 c popcorn**
**3 c boiling water**
**¾ tsp salt**
**3 c milk, reconstituted**

1. Stir popped popcorn into boiling water. Add salt. Boil and stir 5 minutes.
2. Remove from heat, cover pan, let stand several minutes before serving.
3. Serve with warm milk. May add brown sugar, white sugar, cinnamon, raisins, fruit (fresh or canned).

**Overnight Method**: Use cool water instead of boiling. Let soak overnight. In morning, add 3 c milk and warm about 5 minutes on medium heat. Do not boil. Serve with additional milk and any optional ingredients.

---

Per serving (excluding unknown items): 277.4 Calories; 15.4g Fat (48.7% calories from fat); 9.0g Protein; 27.4g Carbohydrate; 25mg Cholesterol; 786mg Sodium. Exchanges: 1 ½ Grain(Starch); ½ Non-Fat Milk; 3 Fat.

# Potato Pancakes

Makes 6 servings. Prep Time: 20 minutes

**6 med potatoes, or 2 lb**
**1 each egg, reconstituted**
**¼ c onion, finely chopped**
**3 tbsp flour**
**1 tsp salt**
**¼ c shortening, or oil**

1. Scrub and wash potatoes. Leave skins on if possible. Shred four cups.
2. Beat egg until thick. Add remaining ingredients except shortening and mix thoroughly.
3. Heat shortening in 12-inch skillet over low heat until melted.
4. Shape mixture into 6-8 patties and place in hot shortening.
5. Cook over medium heat, turning once, about 5 minutes or until brown.

*Flour ground from soft white wheat or all-purpose flour will produce best results.*
Serve with applesauce.

---

Per serving (excluding unknown items): 387.5 Calories; 9.8g Fat (22.3% calories from fat); 8.9g Protein; 68.0g Carbohydrate; 35mg Cholesterol; 388mg Sodium. Exchanges: 4 ½ Grain(Starch); 2 Fat.

# Rice Pudding

Makes 4 servings. Prep Time: 1 hour and 10 minutes

**3 c cooked rice**
**1 c milk, reconstituted**
**2 each eggs, reconstituted**
**¼ c sugar**
**½ c raisins**
**dash cinnamon or nutmeg**
**1 tsp vanilla**

1. Combine all ingredients except cinnamon. Pour into ungreased 1 ½-quart casserole.
2. Place casserole in square pan, 9x9x2 inches, on oven rack. Pour very hot water, one inch deep, into pan.
3. Bake in a 350 degree oven, stirring occasionally, until pudding is creamy, about 1 hour.
4. Sprinkle with cinnamon and serve warm.

**Eggless Variation**: Omit reconstituted eggs and reconstituted milk. Replace with 1 ½ c double-strength reconstituted milk; and 2 tbsp cornstarch liquified in 3-4 tbsp water.

---

Per serving (excluding unknown items): 364.8 Calories; 5.0g Fat (12.4% calories from fat); 9.5g Protein; 70.1g Carbohydrate; 116mg Cholesterol; 67mg Sodium. Exchanges: 2 ½ Grain(Starch); ½ Lean Meat; 2 Fruit; ½ Fat; 1 Other Carbohydrates.

# Squash Griddle Cakes

Makes 4 servings. Prep Time: 30 minutes

**2 c winter squash, cooked and mashed**
**½ tsp salt**
**2 tbsp sugar**
**3 tbsp baking powder**
**2 ¼ c milk**
**3 tbsp soy flour**
**2 c whole wheat flour**
**3 tbsp oil, for frying**

1. Mix all ingredients together. Heat oil on griddle.
2. Fry pancakes, turning to brown on both sides. Cook thoroughly on the first side, until it looks fairly dry on top. Then turn over and brown on the other side. This keeps the pancakes from being too soggy in the middle.

*This recipe from the Depression days uses no oil in the batter. Pumpkin would work well in this recipe.*

Per serving (excluding unknown items): 369.3 Calories; 16.2g Fat (37.7% calories from fat); 12.2g Protein; 48.0g Carbohydrate; 19mg Cholesterol; 339mg Sodium. Exchanges: 2 ½ Grain(Starch); ½ Non-Fat Milk; 3 Fat.

# Sweet Cheese Treat

Makes 6 servings. Prep Time: 15 minutes

**1 ¼ c Pressed Easy Cheese**
**¾ c powdered sugar**
**¾ c dried fruit, finely chopped**
**¼ c shortening**

1. Beat cheese until creamy. Shape into 6 cakes, about 2 inches in diameter.
2. Mix together powdered sugar, dried fruit and shortening.
3. Top cakes with portions of this mixture. Heat in a 400 degree oven for about 5-10 minutes.
4. Serve for breakfast or dessert.

Per serving (excluding unknown items): 190.9 Calories; 8.8g Fat (40.2% calories from fat); 5.5g Protein; 23.8g Carbohydrate; 2mg Cholesterol; 6mg Sodium. Exchanges: ½ Lean Meat; ½ Fruit; 1 ½ Fat; 1 Other Carbohydrates.

# Wheat Berry Cereal

Makes 6 servings. Prep Time: 50 minutes

**3 c whole wheat berries**
**6 c water**
**1 ½ tsp salt, optional**

1. Bring all ingredients to a boil in a large saucepan or Dutch oven. Cover and simmer 45 minutes to 2 hours,until wheat kernels begin to pop open.
2. Add honey or sugar, raisins or chopped fruit as desired.

**Overnight Method**: Simmer 20 minutes instead of 45. Remove from heat. Keep covered and place in hot box, as described in Breakfast introduction. Let sit several hours or overnight. Reheat or serve cold in morning.

Leftovers may be used in your favorite stuffing recipe.

Per serving (excluding unknown items): 239.4 Calories; 0.9g Fat (3.3% calories from fat); 8.6g Protein; 53.1g Carbohydrate; 0mg Cholesterol; 552mg Sodium. Exchanges: 3 ½ Grain(Starch).

# Biscuits and Gravy

Makes 6 servings. Prep Time: 25 minutes

**12 each biscuits**
**3 tbsp shortening**
**2 tbsp cornstarch**
**2 c milk, reconstituted**
**¼ tsp salt**
**pepper, to taste**

1. Melt shortening in 1-quart saucepan over medium heat.
2. Stir in cornstarch and salt with a whisk. Continue heating until bubbly.
3. Remove from heat and stir in milk using whisk.
4. Heat to boiling, stirring constantly. Boil and stir one minute or until gravy turns clear. Add pepper and serve over biscuits.

*2 c beef broth may be substituted for milk.*

Per serving (excluding unknown items): 377.9 Calories; 21.8g Fat (51.8% calories from fat); 6.8g Protein; 38.8g Carbohydrate; 16mg Cholesterol; 865mg Sodium. Exchanges: 2 ½ Grain(Starch); 4 ½ Fat.

# SANDWICHES

Most traditional sandwiches use meat or cheese fillings, so sandwiches made from stored foods, which are mostly vegetarian, will be quite different. Any canned meats or fish would, of course, be easy to substitute in our usual recipes, as would "good old" peanut butter.

We have to look at foods from other lands to find vegetarian-type sandwiches, from the Middle East, India and South-of-the-Border. Some of these have been included. Other recipes have been created using American foods in new, creative ways. Either way, you are in for a treat with these sandwich ideas. No nitrates in these babies, just good, healthy food!

# Baked Bean Sandwich

Makes 4 servings. Prep Time: 10 minutes

**2 c baked beans, leftovers**
**8 slices English muffins**
**8 slices tomato**
**parsley**

1. Toast bread, if desired
2. Spread ¼ c beans on each slice of bread to make 8 open face sandwiches.
3. Garnish with fresh or dried parsley.

*Any bread may be used, such as rye bread, hamburger buns, Boston brown bread.*

Per serving (excluding unknown items): 437.7 Calories; 3.5g Fat (6.8% calories from fat); 17.0g Protein; 89.9g Carbohydrate; 0mg Cholesterol; 1055mg Sodium. Exchanges: 5 Grain(Starch); 2 Vegetable; ½ Fat.

# Barbecue Sandwiches

Makes 4 servings. Prep Time: 30 minutes

**½ can catsup**
**¼ c vinegar**
**2 tbsp chopped onion**
**1 tbsp Worcestershire sauce**
**2 tsp brown sugar**
**¼ tsp garlic powder**
**1 c textured soy protein, chunks**

1. Soak TVP chunks in 1 c hot water until about half-way hydrated. Drain off excess water.
2. In saucepan, heat remaining ingredients to boiling over medium heat.
3. Add TVP and simmer uncovered, stirring occasionally, for 15 minutes.
4. Serve in buns of your choice.

Per serving (excluding unknown items): 194.4 Calories; 0.5g Fat (1.9% calories from fat); 38.7g Protein; 17.0g Carbohydrate; 0mg Cholesterol; 66mg Sodium. Exchanges: 1 Grain(Starch); 5 Lean Meat; ½ Other Carbohydrates.

# Basic Bean Burrito

Makes 1 serving. Prep Time: 15 minutes

**½ c refried beans**
**1 each tortillas**
**1 tbsp chopped green chiles, optional**
**2 tbsp chopped tomato**
**1 tbsp chopped green onions**
**¼ c shredded lettuce**

1. Heat beans and tortilla in steamer (see instructions for reheating leftovers in Stovetop Breads chapter.)
2. Stir chilies into beans.
3. Spread beans on tortilla and sprinkle with all vegetables.
4. Fold or roll and eat while warm.

Per serving (excluding unknown items): 258.4 Calories; 4.0g Fat (13.7% calories from fat); 11.4g Protein; 44.8g Carbohydrate; 0mg Cholesterol; 708mg Sodium. Exchanges: 3 Grain(Starch); ½ Lean Meat; ½ Vegetable; ½ Fat.

# Beefed-up Sandwich Filling

Makes 4 servings. Prep Time: 10 minutes

**½ c textured soy protein**
**1 tsp beef bouillon granules**
**½ c hot water**
**2 tbsp onion, finely chopped**
**2 tbsp sweet pickle relish**
**½ tsp salt**
**1/8 tsp pepper**
**1/3 c sprouts**
**1/3 c mayonnaise-type salad dressing**
**8 slices whole wheat bread**

1. Combine bouillon and hot water in a saucepan. Add TVP and hydrate.
2. Mix in remaining ingredients. Fills 4 sandwiches.

Per serving (excluding unknown items): 386.0 Calories; 10.5g Fat (22.5% calories from fat); 27.9g Protein; 52.9g Carbohydrate; 5mg Cholesterol; 1063mg Sodium. Exchanges: 3 Grain(Starch); 2 ½ Lean Meat; ½ Fruit; 2 Fat.

# East Indian Cheese and Vegetables

Makes 5 servings. Prep Time: 30 minutes

---

**1 ¼ c Pressed Easy Cheese**
**2 tbsp oil**
**1 lg onion, finely chopped**
**1 lg tomato**
**1 tbsp cilantro, if available**
**1/8 tsp pepper**
**1 c whey**
**salt, to taste**
**½ c peas**
**10 lg tortillas**

1. Cut cheese into ½-inch cubes and set aside.
2. In a large skillet, heat oil over medium-high heat. Add onion and cook, stirring, until onion browns lightly. Add tomato, cilantro (or parsley) and pepper. Cook and stir until tomato is very soft, about 5 minutes.
3. Stir in whey and boil over high heat until liquid is almost gone, about 3 minutes. Add salt. Gently stir in peas and cheese cubes. Reduce heat to low and cook, covered, to warm cheese, about 3 minutes.
4. Serve with tortillas, or spoon sauce into 5 pita breads.

*This dish contains traditional Indian seasonings. If these are unavailable, or it they don't suite your taste, try preparing the dish without any extra spices and see how you like it—then try adding some of your own.*

---

Per serving (excluding unknown items): 420.3 Calories; 11.0g Fat (23.6% calories from fat); 16.9g Protein; 63.5g Carbohydrate; 4mg Cholesterol; 603mg Sodium. Exchanges: 3 Grain(Starch); 2 ½ Lean Meat; ½ Vegetable; 2 Fat; 1 Other Carbohydrates.

# East Indian Cheese With Spinach

Makes 5 servings. Prep Time: 30 minutes

**1 ¼ c Pressed Easy Cheese**
**1 lb spinach**
**2 tbsp oil**
**½ tsp cumin**
**1 lg onion, finely chopped**
**½ tsp garlic powder**
**1 lg tomato, chopped**
**1 med jalapeno, optional**
**1 tsp coriander**
**½ tsp turmeric**
**¼ tsp cayenne**
**1 c whey**
**salt, to taste**
**10 lg tortillas**

1. Cut cheese into ½-inch cubes and set aside.
2. Remove stems from spinach; rinse leaves well. Place leaves in a 3-quart saucepan, cover and cook over medium heat until spinach is wilted, stirring occasionally, about 7minutes. Finely chop or puree.
3. In a large skillet, heat oil over medium-high heat, stirring. Add cumin, onion, garlic powder and cook, stirring, until onions brown lightly, about 10 minutes.
4. Stir in tomato, finely chopped jalapeno, coriander, turmeric and cayenne. Cook and stir the sauce until tomato is very soft, about 10 minutes. Add whey and boil over high heat, stirring, until sauce looks well-blended, 3-5 minutes. Season to taste with salt.
5. Reduce heat to low. Stir in spinach and gently mix in cheese cubes. Cook, uncovered, just enough to heat the cheese, about 1 minute. Serve immediately to keep bright spinach color.
6. Serve with tortillas or spoon mixture into 5 pocket breads. Garnish with tomato slices or sprouts.

*This sandwich filling contains traditional Indian seasonings. If these are unavailable, or if they don't suit your taste, try preparing it without any added seasonings and see how you like it—then try adding some of your own.*

---

Per serving (excluding unknown items): 433.8 Calories; 11.4g Fat (23.2% calories from fat); 18.9g Protein; 65.7g Carbohydrate; 4mg Cholesterol; 674mg Sodium. Exchanges: 2 ½ Grain(Starch); 2 ½ Lean Meat; 1 ½ Vegetable; 2 Fat; 1 Other Carbohydrates.

# Middle East Sandwiches

Makes 4 servings. Prep Time: 15 minutes

**2 c cooked garbanzo beans**
**4 tbsp Oriental dressing, see recipe**
**1 c spinach leaves, torn**
**1 lg tomato, chopped**
**4 each pita bread rounds**
**1 c sprouts**

1. Toss together all ingredients except bread.
2. Cut pita bread in half and open pocket gently. Fill with bean mixture.
3. Top with sprouts, if desired. Two halves makes one sandwich.

**Alternate Filling**: ¼ c oil, 4 tsp vinegar, ¼ tsp salt, ¼ tsp garlic powder, ¼ tsp oregano, 2 c cooked garbanzo beans, 1 ½ c thinly sliced and halved zucchini. Mix together and let stand for an hour to blend flavors. Stir again before filling sandwiches.

Per serving (excluding unknown items): 311.4 Calories; 3.1g Fat (8.7% calories from fat); 13.7g Protein; 58.1g Carbohydrate; 0mg Cholesterol; 342mg Sodium. Exchanges: 3 ½ Grain(Starch); ½ Lean Meat; ½ Vegetable.

# Mini Pizza

Makes 1 serving. Prep Time: 10 minutes

**1 each pita bread**
**2 tbsp spaghetti sauce**
**2 tbsp green pepper, finely chopped**
**2 tbsp green onion, finely chopped**
**2 tbsp textured soy protein, soaked**

1. Preheat oven to 425 degrees. Spread sauce on pita bread, or on English muffin halves, if preferred.
2. Sprinkle with vegetables and TVP.
3. Place on ungreased pizza pan or cookie sheet. Bake 5-10 minutes, until heated and crisp.

Per serving (excluding unknown items): 294.4 Calories; 2.5g Fat (7.2% calories from fat); 25.6g Protein; 46.2g Carbohydrate; 0mg Cholesterol; 481mg Sodium. Exchanges: 2 ½ Grain(Starch); 2 ½ Lean Meat; 1 ½ Vegetable; ½ Fat.

# Pita with Falafel Filling

Makes 4 servings. Prep Time: 45 minutes

**½ c chopped onion**
**¼ c parsley**
**2 tbsp sesame seeds, optional**
**1 tbsp oil**
**½ tsp oregano**
**½ tsp mint, dried**
**½ tsp paprika**
**1/8 tsp garlic powder**
**1/8 tsp allspice**
**1/8 tsp cayenne**
**2 c cooked garbanzo beans**
**½ c mayonnaise-type salad dressing**
**4 each pita bread rounds**
**2 c shredded lettuce**
**8 slices tomato**
**2 c sprouts**
**½ c chopped green bell pepper**

1. Make filling by cooking onion in oil over medium heat until almost tender. Add sesame seed and seasonings and cook 1-2 more minutes.
2. Stir in beans and mayonnaise (soft cheese may be substituted.)
3. Puree mixture through food mill.
4. Spoon 3 tbsp filling and some of each of the vegetables into each pita half. for serving.

Per serving (excluding unknown items): 548.1 Calories; 19.5g Fat (30.8% calories from fat); 17.6g Protein; 80.6g Carbohydrate; 8mg Cholesterol; 569mg Sodium. Exchanges: 4 Grain(Starch); ½ Lean Meat; 3 Vegetable; ½ Fruit; 3 Fat.

# PB and What? Sandwich

Makes 1 serving. Prep Time: 5 minutes

**2 slices bread**
**2 tbsp peanut butter**

In place of the traditional "J" (jam or jelly), any of the following may be used as a sandwich filling with peanut butter: thin apple slices, raisins or other chopped dried fruit, applesauce or apple butter, sweet pickle relish, shredded carrots, sprouts, honey or cooked lentils.

Per serving (excluding unknown items): 323.1 Calories; 17.9g Fat (48.1% calories from fat); 12.0g Protein; 31.4g Carbohydrate; 1mg Cholesterol; 423mg Sodium. Exchanges: 2 Grain(Starch); 1 Lean Meat; 3 Fat.

# Sloppy Joes

Makes 5 servings. Prep Time: 30 minutes

**2 tbsp shortening, or oil**
**1 c textured soy protein**
**1 c beef broth**
**½ c chopped onion**
**⅓ c Swiss chard, chopped stalks**
**⅓ c chopped green bell pepper**
**⅓ c catsup**
**¼ c water**
**1 tbsp Worcestershire sauce**
**½ tsp salt**
**5 each hamburger buns, split**

1. Toast hamburger buns.
2. Cook onion, chard and green pepper in shortening in skillet until tender.
3. Heat TVP in beef broth in small saucepan until hydrated.
4. Add TVP to vegetables and stir, cooking about 3 minutes.
5. Add remaining ingredients except buns. Cover and cook over low heat 10-15 minutes.
6. Fill buns with mixture and top with bun half.

**Not-Sloppy Joes**: Sauce may be thickened by adding 1 tbsp cornstarch dissolved in 2 tbsp water during the low heat cooking time. Cook until cornstarch clears.

**Beany Joes**: Substitute 2 c cooked lentils, split peas or beans for the TVP and broth.

**Sloppy Jose**: Use prepared chili (may be thickened, as above) instead of above filling. Or use above filling, but add 1 tsp chili powder and ½ tsp cumin.

*½ c tomato sauce may be substituted for catsup and water. Adjust seasonings to taste. May add sugar, dry mustard, celery seed and paprika.*

---

Per serving (excluding unknown items): 348.7 Calories; 7.8g Fat (18.7% calories from fat); 37.2g Protein; 38.8g Carbohydrate; 0mg Cholesterol; 944mg Sodium. Exchanges: 2 Grain(Starch); 4 ½ Lean Meat; ½ Vegetable; 1 ½ Fat; ½ Other Carbohydrates.

# South-of-the-Border Grilled Sandwiches

Makes 6 servings. Prep Time: 30 minutes

**2 c refried beans**
**1 tbsp shortening**
**1 tbsp finely chopped onion**
**2 tbsp green pepper, finely chopped**
**1 tsp chili powder**
**12 slices whole wheat bread**
**6 tbsp shortening**
**½ c green chiles, cut in strips**

1. In skillet, cook onion and green pepper in 1 tbsp shortening until tender.
2. Stir in refried beans and chili powder. Cook over low heat until warmed. Remove from pan.
3. Using 6 tbsp of shortening, "butter" one side of each piece of bread. Place buttered side of 6 slices down on skillet over medium heat.
4. Spread ⅓ c bean mixture on each piece of bread. Top with chili strips and remaining slices of bread, buttered side up.
5. Grill sandwiches on each side until brown. Serve warm.

Per serving (excluding unknown items): 431.5 Calories; 19.4g Fat (39.0% calories from fat); 13.5g Protein; 54.8g Carbohydrate; 0mg Cholesterol; 805mg Sodium. Exchanges: 3 ½ Grain(Starch); 3 ½ Fat.

# Boston Bean Sandwiches

Makes 5 servings. Prep Time: 15 minutes

**2 c baked beans**
**2 tbsp finely chopped onion**
**2 tbsp chili sauce, optional**
**4 tbsp mayonnaise-type salad dressing**
**1 16 oz Boston brown bread, canned**

1. Slice bread loaf into 10 slices. Spread with mayonnaise.
2. Stir together beans, onion, chili sauce; Warm in saucepan.
3. Divide equally among slices of bread and serve as open face sandwiches.

Per serving (excluding unknown items): 160.6 Calories; 4.5g Fat (23.2% calories from fat); 5.5g Protein; 28.2g Carbohydrate; 3mg Cholesterol; 545mg Sodium. Exchanges: 1 ½ Grain(Starch); 1 Fat.

# Spiced Cheese Spread

Makes 6 servings. Prep Time: 50 minutes

**2 qt milk, reconstituted**
**¼ tsp nutmeg**
**¼ tsp cinnamon**
**½ tsp salt**

1. Make Drained (soft) cheese using Easy Cheese recipe, but add above spices along with the salt during the cooking process.
2. Spread cheese thickly on toasted English muffins or toast and drizzle with honey for breakfast.
3. For lunch, chopped dried or fresh fruit and/or nuts may be added as a sandwich filling.

Per serving (excluding unknown items): 200.6 Calories; 10.9g Fat (48.6% calories from fat); 10.7g Protein; 15.3g Carbohydrate; 44mg Cholesterol; 337mg Sodium. Exchanges: 1 Non-Fat Milk; 2 Fat.

# Mexican Cheese Spread

Makes 6 servings. Prep Time: 50 minutes

**2 qt milk, reconstituted**
**½ tsp salt**
**⅛ tsp cumin**
**dash cayenne**
**¼ tsp chili powder**

1. Make Drained (soft) cheese using Easy Cheese recipe. Add above seasonings along with salt to the heating milk.
2. To serve, mound cheese in a bowl and top with Fresh Salsa. Spread on crackers, tortilla chips or pita chips.

Per serving (excluding unknown items): 200.4 Calories; 10.9g Fat (48.6% calories from fat); 10.7g Protein; 15.2g Carbohydrate; 44mg Cholesterol; 338mg Sodium. Exchanges: 1 Non-Fat Milk; 2 Fat.

# Tuna Salad Sandwich

Makes 4 servings. Prep Time: 10 minutes

**1 can tuna in water, drained**
**¼ c green onion, chopped**
**¼ c Swiss chard, chopped stalks**
**3 tbsp mayonnaise-type salad dressing**
**8 slices whole wheat bread**

1. Blend together all ingredients, except bread.
2. Use as filling for 4 sandwiches.
3. May top with lettuce, sprouts or tomato.

**Tuna-Pickle Sandwich**: Substitute ¼ c pickle relish for chard.
**Hot Tuna Salad Sandwich**: Spread mixture on 4 burger bun halves. Top with other halves. Place in a covered casserole. Bake in a 400 degree preheated oven, 15-20 minutes.
**Tuna-Wheat Sandwich**: Add ½ c cooked whole or cracked wheat. Add mayonnaise as needed to moisten.

Per serving (excluding unknown items): 291.7 Calories; 7.5g Fat (22.2% calories from fat); 17.1g Protein; 41.9g Carbohydrate; 13mg Cholesterol; 643mg Sodium. Exchanges: 2 ½ Grain(Starch); 1 Lean Meat; 1 ½ Fat.

# Under Wraps

Makes 1 serving. Prep Time: 15 minutes

**warmed tortilla**
**any cooked bean or TVP, warmed**
**any sauteed or cooked vegetable**
**seasonings and dressing as desired**

1. Warm tortilla and beans in steamer (see instructions for reheating leftovers in Stovetop Breads chapter.)
2. Heat or saute vegetables as desired. Sauteed onion and green pepper are good choices.
3. Fill half of tortilla with portion of beans and vegetables.
4. Top as desired: any dressing, herb, spice, salsa, etc.
5. Wrap tortilla in half and enjoy while warm.

Per serving (excluding unknown items): 0.0 Calories; 0.0g Fat (0.0% calories from fat); 0.0g Protein; 0.0g Carbohydrate; 0.0mg Cholesterol; 0.0mg Sodium. Exchanges: Free.

# Vegetarian BLT Sandwich

Makes 1 serving. Prep Time: 5 minutes

**2 slices bread**
**1 tbsp mayonnaise-type salad dressing**
**2 each lettuce leaves**
**3 slices tomato**

1. The BLT stands for BREAD, lettuce and tomato! Serve with a glass of milk or a serving of baked beans for complementary proteins.
2. Spread bread slices with mayonnaise. Layer lettuce, tomato and top with other slice of bread.

**Sprout Sandwich**: Substitute sprouts for the lettuce.

Per serving (excluding unknown items): 270.1 Calories; 8.0g Fat (25.2% calories from fat); 7.5g Protein; 45.7g Carbohydrate; 4mg Cholesterol; 408mg Sodium. Exchanges: 1 ½ Grain(Starch); 3 ½ Vegetable; 1 ½ Fat.

# Veggie Burgers

Makes 6 servings. Prep Time: 1 hour

**2 c lentil, bean, or rice mixture for patties (see recipe)**
**6 each hamburger buns**

1. Prepare lentil, bean or rice patties as directed in Stovetop Main Dishes.
2. Make Whole Wheat Bread dough. Using half, form into about 6 hamburger bun shapes and bake as directed. Split in half when cooled.
3. Place cooked patties in buns. Season and garnish as you would traditional hamburgers: catsup, mayo, mustard, Worcestershire sauce, tomato slices, lettuce, sprouts, pickles, etc.

Per serving (excluding unknown items): 123.0 Calories; 2.2g Fat (16.3% calories from fat); 3.7g Protein; 21.6g Carbohydrate; 0mg Cholesterol; 241mg Sodium. Exchanges: 1 ½ Grain(Starch); ½ Fat.

# VEGETABLES AND SALADS

Recipes for vegetables and vegetable salads do not require much adjustment for preparation without electrical appliances, since they are most often prepared on top of the stove. The reader may refer to regular cookbooks or personal cooking experience for information regarding preparation of all the various vegetables. What is included in this section are recipes of a more unusual nature or recipes that lend themselves to an "off the grid" viewpoint. Main dish salads will also have considerable focus.

One preparation detail that *will* change off the grid is the possibility of using butter to season cooked vegetables. Since butter or margarine is not available, other seasonings must be considered. If a fat source is desired, olive oil, seasoned oils or dressings are good alternatives to butter. This is customary in the cooking of southern Europe, for example. Medium White Sauce may also be used for seasoning and variety. Another choice is to use herbs and salt to give flavor to vegetable dishes. Be creative and try new combinations! The following recipes offer some examples.

Vegetables may be used fresh, canned, dried or fresh from the root cellar. It is recommended that there is some of each type to ensure that vegetables are always available for health and variety in the diet. Books on gardening, canning, drying and root cellaring would help in this process. The Cooperative Extension Service, local library and Internet are all good places to go for reliable information.

Another form of gardening—sprouting—would be very useful to provide fresh greens during the winter months, since it can be done indoors. Seeds and legumes such as wheat, rye, radishes, mung beans, alfalfa, lentils, white beans, pinto beans, etc. can all be used. Be cautioned that sprouted legumes should be stir fried or steamed two minutes before eating to inactivate an "anti-protein" substance in raw legumes. Again, the reader is encouraged to seek out further information on sprouting, as listed above.

# Autumn Carrots

Makes 6 servings. Prep Time: 15 minutes

**3 c shredded carrots**
**2 lg apples, coarsely chopped**
**2 tbsp chopped nuts, optional**
**½ tsp coriander**
**1 tbsp oil**

1. In a large skillet, heat oil.
2. Add remaining ingredients and stir fry until tender-crisp, about 5 minutes.
3. Serve while hot.

Per serving (excluding unknown items): 89.1 Calories; 4.2g Fat (39.8% calories from fat); 1.2g Protein; 13.2g Carbohydrate; 0mg Cholesterol; 20mg Sodium. Exchanges: 1 Vegetable; ½ Fruit; 1 Fat.

# Baked Whole Onions

Makes 6 servings. Prep Time: 50 minutes

**6 med onions, unpeeled**
**½ tsp garlic powder**
**¼ c olive oil**
**½ tsp salt**
**1 tbsp parsley flakes**

1. Place onions directly on middle oven rack. Bake in a 350 degree oven until onions are tender, 30-40 minutes.
2. Remove skins. Place onions upright in serving dish.
3. Cut each onion into quarters, about halfway through; separate slightly.
4. Heat oil and seasonings in a small saucepan and pour over onions.

Per serving (excluding unknown items): 136.6 Calories; 9.2g Fat (58.5% calories from fat); 1.8g Protein; 12.9g Carbohydrate; 0mg Cholesterol; 182mg Sodium. Exchanges: 2 Vegetable; 2 Fat.

# Cabbage and Linguine

Makes 6 servings. Prep Time: 25 minutes

**12 ozs linguine, uncooked**
**4 tbsp shortening**
**2 tbsp cornstarch**
**½ tsp mace**
**½ tsp salt, or to taste**
**1 c milk, reconstituted**
**4 c shredded cabbage**
**2 tbsp parsley**

1. Prepare linguine (or spaghetti) as directed on package. Drain and place in mixing bowl. It may be tosses with a small amount of oil to keep it from sticking together.
2. Meanwhile, in large skillet, melt shortening. Stir in cornstarch, salt, mace.
3. Blend in milk with a whisk. Cook and stir until thickened.
4. Add cabbage and parsley. Cover and cook on low until cabbage is tender-crisp. Serve over hot linguine.

Per serving (excluding unknown items): 754.9 Calories; 12.8g Fat (15.4% calories from fat); 23.9g Protein; 134.1g Carbohydrate; 6mg Cholesterol; 220mg Sodium. Exchanges: 8 ½ Grain(Starch); ½ Vegetable; 2 Fat.

# Potato Planks

Makes 6 servings. Prep Time: 1 hour

**6 lg baking potatoes**
**¼ c oil**
**salt and pepper, to taste**

1. Scrub potatoes. Cut into 8 wedge-shaped "planks" lengthwise.
2. Coat with oil (may be sprayed with pan spray instead.)
3. Sprinkle with salt, pepper and any other seasonings, as desired: garlic, paprika, chili powder, etc.
4. Bake in a 425 degree oven about 20 minutes, until golden brown on top. Turn over and bake and additional 20 minutes more until done on all sides.

Per serving (excluding unknown items): 198.8 Calories; 9.2g Fat (40.9% calories from fat); 3.1g Protein; 27.0g Carbohydrate; 0mg Cholesterol; 9mg Sodium. Exchanges: 1 ½ Grain(Starch); 2 Fat.

# Cabbage Wedges

Makes 6 servings. Prep Time: 30 minutes

**1 sm cabbage head, cut in 6 wedges**
**1 tbsp sugar**
**3 tbsp vinegar**
**1 tbsp oil**
**1 tsp finely chopped onion**
**¼ tsp salt**
**¼ tsp celery seed**
**¼ tsp ginger**

1. Heat 1-inch salted water (1/2 tsp salt to 1 c water) to boiling in a 10-inch skillet. Add cabbage.
2. Cover and heat to boiling. Reduce heat, and simmer until tender-crisp, about 15 minutes.
3. Remove to serving platter with slotted spoon. Discard salted water.
4. Add remaining ingredients to skillet and heat, stirring occasionally. Pour sauce over cabbage wedges.

**Italian Cabbage**: Substitute 4 tbsp Italian Salad Dressing for the above sauce.

**Creamed Cabbage**: Substitute ½ c white sauce for the above sauce.

Per serving (excluding unknown items): 70.4 Calories; 2.8g Fat (32.6% calories from fat); 2.2g Protein; 10.9g Carbohydrate; 0mg Cholesterol; 118mg Sodium.
Exchanges: 1 ½ Vegetable; ½ Fat.

## Carrot Casserole

Makes 6 servings. Prep Time: 1 hour

**4 ½ c shredded carrots**
**1 lg chopped onion**
**1 tsp salt**
**¼ tsp dry mustard**
**½ tea marjoram**
**¾ c hot water**
**¼ c chopped nuts, optional**

1. Mix carrots and seasonings in an ungreased 1 ½-quart casserole.
2. Pour water over carrots. Cover and cook in a 350 degree oven until tender, about 45 minutes.
3. Sprinkle with nuts before serving, if desired.

*This recipe could be easily changed to stovetop by simmering in a large skillet.*

---

Per serving (excluding unknown items): 82.0 Calories; 3.6g Fat (36.3% calories from fat); 2.2g Protein; 11.9g Carbohydrate; 0mg Cholesterol; 387mg Sodium. Exchanges: 2 Vegetable; ½ Fat.

## Potato Poppers

Makes 4 servings. Prep Time: 45 minutes

**½ c finely chopped onion**
**1 tbsp shortening**
**½ c mashed potatoes**
**1 c cooked rice**
**1 tbsp tomato paste, or catsup**
**½ tsp salt**
**½ c bread crumbs**
**dash pepper**

1. Preheat oven to 350 degrees. Saute onion in shortening until tender.
2. Combine all ingredients and form into 1 ½-inch balls.
3. Bake on a lightly greased cookie sheet until lightly browned, about 15-20 minutes. Or poppers may be fried in 3 tbsp. oil in a skillet until brown.

---

Per serving (excluding unknown items): 177.4 Calories; 4.6g Fat (23.5% calories from fat); 3.9g Protein; 29.9g Carbohydrate; 1mg Cholesterol; 476mg Sodium. Exchanges: 2 Grain(Starch); ½ Vegetable; 1 Fat.

# Hot Whole Wheat Salad

Makes 6 servings. Prep Time: 1 hour and 15 minutes

**1 c whole wheat berries**
**4 c water**
**½ c chopped onion**
**2 tbsp oil**
**3 c summer squash, sliced**
**1 c shredded carrots**
**1 c peas, fresh**
**⅓ c water**
**2 tsp beef bouillon granules**
**½ tsp basil**

1. Wash whole wheat by rinsing in a strainer. Heat berries and 4 c water in a 3-quart saucepan. Boil 2 minutes. Remove from heat. Cover and let stand 1 hour. Do not drain.
2. Heat to boiling. Reduce heat, cover and simmer until tender and popped open, about 45-60 minutes.
3. Cook and stir onion and squash in oil over medium heat in a 10-inch skillet until tender, about 5 minutes.
4. Stir in wheat and remaining ingredients. Heat to boiling. Reduce heat, cover and simmer until liquid is absorbed, about 10 minutes.

Per serving (excluding unknown items): 168.6 Calories; 5.3g Fat (26.4% calories from fat); 5.5g Protein; 27.4g Carbohydrate; 0mg Cholesterol; 217mg Sodium. Exchanges: 1 ½ Grain(Starch); 1 Vegetable; 1 Fat.

# Italian Potatoes

Makes 6 servings. Prep Time: 30 minutes

**6 tbsp Italian Salad Dressing**
**4 tbsp green onion, sliced**
**2 tbsp parsley, fresh**
**¼ tsp fennel seed, crushed**
**dash pepper**
**4 lg potatoes, sliced ¼-inch thick**

1. In a large skillet, over medium-low heat, cook all ingredients, stirring to coat potatoes.
2. Cover and cook slowly, 15-20 minutes, until potatoes are tender.

Per serving (excluding unknown items): 260.0 Calories; 7.4g Fat (24.8% calories from fat); 5.2g Protein; 45.0g Carbohydrate; 0mg Cholesterol; 133mg Sodium. Exchanges: 3 Grain(Starch); 1 ½ Fat.

# Lettuce Skillet

Makes 6 servings. Prep Time: 25 minutes

**3 c summer squash, sliced**
**1 lg onion, thinly sliced**
**½ tsp garlic powder**
**2 tbsp oil**
**1 tbsp vinegar**
**½ tsp salt**
**½ tsp dry mustard**
**1/8 tsp pepper**
**1 lb leaf lettuce, coarsely shredded**

1. Cook and stir squash, onion and oil in a 10-inch skillet over medium heat until vegetables are tender.
2. Stir in vinegar and seasonings. Add lettuce (should be about 6 c.)
3. Cover and simmer just until lettuce is wilted, about 5 minutes.
4. Toss. Serve immediately.

Per serving (excluding unknown items): 73.7 Calories; 4.9g Fat (55.9% calories from fat); 1.9g Protein; 6.8g Carbohydrate; 0mg Cholesterol; 186mg Sodium. Exchanges: 1 Vegetable; 1 Fat.

# Ratatouille (Eggplant Stew)

Makes 6 servings. Prep Time: 1 hour and 30 minutes

**2 ½ c eggplant, cut in ½" cubes**
**1 c thinly sliced onion**
**1 tsp garlic powder**
**2 tbsp oil**
**4 med green bell peppers, sliced**
**2 c tomatoes, quartered**
**3 c zucchini, sliced ½-inch thick**
**1 tsp oregano**
**½ tsp salt**
**pepper, to taste**

1. Cook onions in oil in a 6-quart pot. Add peppers, garlic, tomatoes and zucchini. Cook until heated.
2. Add eggplant and seasonings. Cover and cook very slowly for about one hour.
3. Uncover and cook 15 minutes longer, to thicken stew.

*1 pint canned tomatoes may be substituted for the fresh tomatoes. This traditional French dish is richly flavored and tastes like a full meal, although it is only vegetables.*

Serve with French bread and milk for a complete meal.

---

Per serving (excluding unknown items): 99.6 Calories; 5.0g Fat (40.7% calories from fat); 2.5g Protein; 13.7g Carbohydrate; 0mg Cholesterol; 188mg Sodium. Exchanges: 2 Vegetable; 1 Fat.

# Mashed Potatoes

Makes 6 servings. Prep Time: 35 minutes

**6 med potatoes**
**water for cooking**
**1/2 tsp salt**
**dash pepper**
**½ c milk, reconstituted**
**2 tbsp oil**

1. Wash, pare and cut potatoes into chunks.
2. Boil in 1 inch salted water until tender, about 25 minutes. Drain.
3. Mash completely. Beat in enough milk to make smooth and fluffy.
4. Add salt and pepper. Add oil, if desired.
5. Garnish with paprika or chives, if desired.

Per serving (excluding unknown items): 408.6 Calories; 9.8g Fat (21.2% calories from fat); 10.5g Protein; 71.7g Carbohydrate; 12mg Cholesterol; 422mg Sodium. Exchanges: 4 ½ Grain(Starch); ½ Non-Fat Milk; 2 Fat.

# Skillet Fries

Makes 6 servings. Prep Time: 50 minutes

**6 med potatoes**
**2 tbsp shortening**
**1 lg onion, thinly sliced**
**1 ½ tsp salt**
**pepper, to taste**
**2 tbsp shortening**

1. Prepare potatoes as directed for Scalloped Potatoes.
2. Heat shortening in a heavy skillet until melted. Layer ⅓ each of the potato and onion; sprinkle with ½ tsp salt and dash pepper. Repeat twice.
3. Dot top layer with additional 2 tbsp shortening.
4. Cook, covered, over medium heat 20 minutes.
5. Uncover and cook, turning once when potatoes are browned on bottom.

*The secret is to not turn the potatoes before they are well-browned.*

Per serving (excluding unknown items): 367.7 Calories; 8.9g Fat (21.3% calories from fat); 7.7g Protein; 66.5g Carbohydrate; 0mg Cholesterol; 555mg Sodium. Exchanges: 4 ½ Grain(Starch); ½ Vegetable; 1 ½ Fat.

## Steamed Radishes

Makes 6 servings. Prep Time: 20 minutes

**24 med radishes, cleaned**
**2 tbsp oil**
**¼ tsp thyme**
**¼ tsp basil**
**¼ tsp fennel seed**
**dash pepper**

1. Place radishes in saucepan. Cover with salted water.
2. Simmer over low heat, covered, until tender, 10-12 minutes. Drain.
3. Add oil and seasonings. Stir over medium heat until radishes are well coated.
4. Serve immediately.

Per serving (excluding unknown items): 43.8 Calories; 4.6g Fat (92.0% calories from fat); 0.2g Protein; 0.8g Carbohydrate; 0mg Cholesterol; 4mg Sodium. Exchanges: 1 Fat.

## Sauteed Greens

Makes 6 servings. Prep Time: 20 minutes

**1 lb Swiss chard**
**½ tsp garlic powder**
**2 tbsp olive oil**
**salt, to taste**

1. Any greens such as kale, spinach, escarole, etc may be substituted for the Swiss chard. Thoroughly wash and chop greens. It should make 5-6 c.
2. Heat oil in large skillet. Add greens and saute, stirring frequently until greens are tender.
3. Add garlic and salt about halfway through the cooking time.

Per serving (excluding unknown items): 53.8 Calories; 4.7g Fat (72.1% calories from fat); 1.3g Protein; 2.8g Carbohydrate; 0mg Cholesterol; 148mg Sodium. Exchanges: ½ Vegetable; 1 Fat.

# Bean Bonanza

Makes 10 servings. Prep Time: 15 minutes

**1 pt green beans, canned**
**2 c kidney beans, cooked**
**2 c garbanzo beans, cooked**
**1 pt wax beans, canned**
**1 c chopped green bell pepper**
**½ c chopped onion**
**½ c oil**
**½ c vinegar**
**¾ c sugar**
**1 tsp salt**
**¼ tsp garlic powder**
**½ tsp pepper**

1. Combine all vegetable ingredients in a large bowl.
2. Combine dressing ingredients in a jar with a tight-fitting lid. Shake vigorously 1 minutes. Or, whisk together in a small bowl.
3. Pour over salad and toss well. Let stand before serving.

*This is a variation of a three-bean salad—it has FOUR beans, instead!*

Per serving (excluding unknown items): 277.9 Calories; 12.1g Fat (37.8% calories from fat); 7.0g Protein; 37.6g Carbohydrate; 0mg Cholesterol; 514mg Sodium. Exchanges: 1 Grain(Starch); 1 Vegetable; 1 Fruit; 2 ½ Fat; 1 Other Carbohydrates.

# Black Bean and Rice Salad

Makes 6 servings. Prep Time: 15 minutes

**2 c cooked rice**
**2 c black beans, cooked**
**½ c Italian Salad Dressing**
**¼ c chopped green bell pepper**
**1/8 tsp pepper**
**1 med tomato, chopped**
**2 tbsp chopped cilantro, optional**

1. Be sure rice and beans are completely cooled. Combine all ingredients in a large bowl.
2. Stir thoroughly to combine flavors.

Per serving (excluding unknown items): 254.5 Calories; 10.0g Fat (35.1% calories from fat); 7.2g Protein; 34.5g Carbohydrate; 0mg Cholesterol; 158mg Sodium. Exchanges: 2 Grain(Starch); 2 Fat.

# Chili Bean Salad

Makes 10 servings. Prep Time: 20 minutes

**2 c kidney beans, cooked**
**2 c pinto beans, cooked**
**2 c garbanzo beans, cooked**
**2 c canned corn**
**½ c chopped green onions**
**2 tbsp parsley flakes**
**½ c green chilies, diced**
**2 tbsp oil**
**¼ c vinegar**
**½ tsp garlic powder**
**1 tsp chili powder**
**¼ tsp cumin**
**salt and pepper, to taste**

1. Combine all beans and vegetables in a large bowl.
2. Mix dressing ingredients and pour over salad.
3. Let sit several hours to combine flavors.

*Add tortillas, tortilla chips or fry bread to complete the meal.Or serve over rice.*

---

Per serving (excluding unknown items): 206.4 Calories; 4.3g Fat (17.7% calories from fat); 9.9g Protein; 34.7g Carbohydrate; 0mg Cholesterol; 116mg Sodium. Exchanges: 2 Grain(Starch); ½ Lean Meat; 1 Fat.

# Sauerkraut Salad

Makes 6 servings. Prep Time: 15 minutes

**1 qt sauerkraut**
**1 c sugar**
**½ c oil**
**¼ c vinegar**
**1 med onion, chopped**
**1 med green bell pepper, chopped**
**1 med red bell pepper, chopped**
**1 tbsp caraway seed**

1. Rinse and drain sauerkraut thoroughly.
2. Combine sugar, oil, vinegar and caraway seed and whisk until blended.
3. Add vegetables and toss well.

**Bean-Sauerkraut Salad**: Add 2 c cooked garbanzo beans or 2 c cooked kidney beans. Omit caraway seed.

*The red and green bell peppers, although not essential, add a striking color contrast.*

---

Per serving (excluding unknown items): 340.6 Calories; 18.6g Fat (47.1% calories from fat); 2.2g Protein; 44.9g Carbohydrate; 0mg Cholesterol; 1042mg Sodium. Exchanges: 2 Vegetable; 2 Fruit; 3 ½ Fat; 2 ½ Other Carbohydrates.

# Lentil Salad

Makes 6 servings. Prep Time: 45 minutes

**1 ½ c lentils**
**4 c water**
**2 med onions, sliced**
**8 whole cloves**
**2 each bay leaves**
**½ tsp garlic powder**
**½ tsp salt**
**½ tsp Tabasco sauce**
**⅓ c oil**
**½ c vinegar**
**½ c parsley, chopped**
**½ c chopped onion**
**2 med tomatoes, cut in wedges**
**2 c sliced cucumbers**
**salad greens**

1. In a large saucepan, combine lentils, water, 2 c onions, cloves, bay leaves, garlic, salt.
2. Bring to a boil. Reduce heat and simmer, covered, 30 minutes or until lentils are tender. Drain, remove bay leaves and cloves. Rinse in cold water.
3. Add remaining ingredients except greens and stir to combine.
4. Serve in bowls line with salad greens.

*Serve this with rolls to complement the protein and complete the meal.*

---

Per serving (excluding unknown items): 347.0 Calories; 14.7g Fat (35.4% calories from fat); 15.8g Protein; 44.5g Carbohydrate; 0mg Cholesterol; 225mg Sodium. Exchanges: 2 Grain(Starch); 1 Lean Meat; 1 ½ Vegetable; 3 Fat.

# Mediterranean Barley Salad

Makes 6 servings. Prep Time: 45 minutes

**1 c pearl barley**
**3 c chicken broth, from base**
**½ tsp garlic powder**
**1 tbsp parsley flakes**
**3 tbsp vinegar**
**2 tbsp olive oil**
**¼ tsp salt**
**⅛ tsp pepper**
**⅛ tsp fennel seed**

1. Rinse barley and place in a 2-quart saucepan with broth. Bring to a boil, reduce heat, cover and simmer until barley is tender, about 35 minutes.
2. Drain barley through a fine sieve; then transfer to serving bowl. Cool.
3. Add remaining ingredients and toss well.

*Shredded carrots, other vegetables or cooked beans may be added to this salad.*

Per serving (excluding unknown items): 197.9 Calories; 6.2g Fat (27.7% calories from fat); 8.9g Protein; 27.5g Carbohydrate; 1mg Cholesterol; 875mg Sodium. Exchanges: 1 ½ Grain(Starch); ½ Lean Meat; 1 Fat.

# Kraut and Apple Salad

Makes 6 servings. Prep Time: 15 minutes

**1 qt sauerkraut**
**2 lg apples, diced**
**½ c finely chopped onion**
**1 lg carrot, shredded**
**1 tbsp parsley flakes**
**2 tbsp caraway seed**
**¼ c oil**
**3 tbsp vinegar**
**1 tsp sugar**
**½ tsp salt**
**½ tsp pepper**

1. Rinse and drain sauerkraut thoroughly.
2. Toss together all ingredients in a large bowl.
3. Blend flavors for 2 hours.

*Red apples with the peel left on make a pretty color contrast in this dish.*

Per serving (excluding unknown items): 170.1 Calories; 10.1g Fat (50.1% calories from fat); 2.3g Protein; 18.5g Carbohydrate; 0mg Cholesterol; 1222mg Sodium. Exchanges: 1 ½ Vegetable; ½ Fruit; 2 Fat.

## Old-Fashioned Cole Slaw

Makes 6 servings. Prep Time: 30 minutes

**½ head cabbage, finely chopped**
**½ sm green bell pepper, chopped**
**⅓ c white vinegar**
**3 tbsp oil**
**2 tbsp sugar**
**1 tsp onion flakes**
**1 tsp salt**
**½ tsp celery**
**½ tsp dry mustard**
**⅛ tsp pepper**

1. Mix all ingredients. Cover and let stand at least an hour to blend flavors. This cole slaw contains no ingredients that are highly perishable, as the traditional mayonnaise dressing would be.

Per serving (excluding unknown items): 103.9 Calories; 7.3g Fat (59.9% calories from fat); 1.2g Protein; 9.8g Carbohydrate; 0mg Cholesterol; 372mg Sodium. Exchanges: 1 Vegetable; ½ Fruit; 1 ½ Fat; ½ Other Carbohydrates.

## Pasta Salads

Makes 6 servings. Prep Time: 30 minutes

**½ lb pasta, any large shape**
**½ c Salad Dressing, your choice**
**3 c chopped fresh vegetables**
**1 c cooked beans, your choice**

1. Cook pasta to "al dente" stage, or slightly underdone. Immediately rinse in cold water to stop cooking process.
2. Stir in dressing, vegetables and beans, as desired. Serve cold

**Tuna Pasta Salad**: Replace the beans with a 6-oz can of tuna.

*The wheat pasta and the beans combine to form complementary proteins, making a full meal in one dish.*

Per serving (excluding unknown items): 0.0 Calories; 0.0g Fat (0.0% calories from fat); 0.0g Protein; 0.0g Carbohydrate; 0.0mg Cholesterol; 0.0mg Sodium. Exchanges: Free.

# Tabbouli

Makes 6 servings. Prep Time: 20 minutes

**3 c cracked wheat cereal, cooked**
**2 lg tomatoes, finely diced**
**6 med green onions, sliced**
**3 tbsp chopped fresh mint**
**1 c parsley, finely chopped**
**3 tbsp olive oil**
**¼ c vinegar**
**¼ c water**
**salt and pepper, to taste**

1. Stir and fluff cracked wheat.
2. Add remaining ingredients and stir well.
3. Let flavors blend at least 2 hours before serving.

*This traditional Middle Eastern dish is usually made with bulgur wheat, basically cracked wheat with some of the bran removed. It cooks up more quickly and stays fluffier than regular cracked wheat.*

Per serving (excluding unknown items): 206.4 Calories; 7.5g Fat (29.7% calories from fat); 6.8g Protein; 33.2g Carbohydrate; 0mg Cholesterol; 31mg Sodium. Exchanges: 1 Grain(Starch); 3 Vegetable; 1 ½ Fat.

# Taco Salad

Makes 6 servings. Prep Time: 30 minutes

**2 c chili**
**1 ½ c salsa, see recipe**
**6 c shredded lettuce**
**2 med tomatoes, diced**
**1 c cucumber slices**
**6 c tortilla chips**

1. Arrange lettuce, chili (any recipe), tomatoes and cucumbers (or any other chopped or shredded vegetable) on individual plates.
2. Just before serving, add chips and pour salsa on top.

*Ranch style dressing may substitute for the traditional sour cream on taco salad.*

Per serving (excluding unknown items): 348.4 Calories; 16.3g Fat (40.8% calories from fat); 10.2g Protein; 45.7g Carbohydrate; 14mg Cholesterol; 845mg Sodium. Exchanges: 3 Grain(Starch); ½ Lean Meat; 1 ½ Vegetable; 3 Fat.

# DRESSINGS AND SAUCES

Sauces are useful for adding flavor, changing texture and giving variety to an otherwise repetitive diet. Some of these are for main dishes or vegetables; others are sweet and intended for topping cakes or desserts. The maple-flavored syrup is especially good for pancakes or fried mush. Check out the tomato sauce recipe and its variations. There is no need to buy seasoned tomato sauce when it is so easy to make your own. Just about all that is needed are the herbs and the tomatoes.

Salad dressings do for fresh vegetables what sauces do for cooked foods. Recipes for dressings are included because without a refrigerator, store-bought dressings, with their "refrigerate after opening" labels, would be unusable. These recipes have been written to make smaller portions so they can be used up on an almost daily basis.

Salad dressings are made from either a vinegar and oil base or a mayonnaise-type base. Included are recipes for making basic mayonnaise and a mayonnaise-type dressing. Mayonnaise is mostly oil emulsified by egg yolk (or soy flour, in this case.) The mayonnaise-type dressing, with similar looks, flavors and uses, is mostly milk thickened with cornstarch. Both usually contain animal foods (milk and eggs), so they are highly susceptible to bacterial growth. Vinegar and oil, on the other hand, do not generally have any perishable ingredients, so they may be kept without refrigeration for a longer period of time. Eventually, however, flavors will deteriorate and the oil will become rancid. Pay close attention to the following recipes to know which are perishable and which are not.

## Catalina Salad Dressing

Makes 8 servings. Prep Time: 10 minutes

**2 tsp salt**
**¼ c catsup**
**2 tsp paprika**
**⅓ c vinegar**
**½ c sugar**
**½ c oil**

1. Combine all ingredients except oil using a whisk.
2. Slowly whisk in oil to blend thoroughly.

*This dressing contains catsup, which may spoil if not refrigerated.*

Per serving (excluding unknown items): 179.5 Calories; 13.7g Fat (66.4% calories from fat); 0.2g Protein; 15.4g Carbohydrate; 0mg Cholesterol; 622mg Sodium. Exchanges: 1 Fruit; 2 ½ Fat; 1 Other Carbohydrates.

## Cheese Dressing

Makes 16 servings. Prep Time: 15 minutes

**1 ¼ c Pressed Easy Cheese**
**1 ¼ c olive oil**
**½ c vinegar**
**¼ c minced onion**
**1 tsp rosemary**

1. Cut cheese into small chunks. Mix with all other ingredients.
2. Use or cover and keep at room temperature up to 1 month. Stir occasionally.
3. Spoon over salad greens. Makes 2 ½ cups dressing.

Per serving (excluding unknown items): 161.0 Calories; 16.9g Fat (92.9% calories from fat); 2.0g Protein; 0.9g Carbohydrate; 1mg Cholesterol; 2mg Sodium. Exchanges: ½ Lean Meat; 3 ½ Fat.

## Cooked Salad Dressing

Makes 8 servings. Prep Time: 20 minutes

**1 tbsp cornstarch**
**1 tbsp sugar**
**½ tsp salt**
**½ tsp dry mustard**
**¾ c milk, reconstituted**
**1 tbsp soy flour**
**3 tbsp vinegar**
**1 ½ tsp oil**

1. Mix cornstarch, sugar, salt, mustard and soy flour in a 2-quart saucepan.
2. Stir in milk gradually, using whisk to avoid lumping.
3. Heat to boiling over med. heat (or use a double boiler), stirring constantly.
4. Boil and stir 1-2 minutes, until thick. Remove from heat.
5. Stir in vinegar and oil (shortening may be substituted.) Cool before using.

*THIS DRESSING WILL SPOIL QUICKLY IF NOT REFRIGERATED BECAUSE IT CONTAINS MILK!*

Per serving (excluding unknown items): 35.5 Calories; 1.8g Fat (43.8% calories from fat); 1.0g Protein; 4.1g Carbohydrate; 3mg Cholesterol; 145mg Sodium.
Exchanges: ½ Fat.

## Ranch Style Dressing

Makes 5 servings. Prep Time: 10 minutes

**⅓ c mayonnaise-type salad dressing**
**¼ c milk, reconstituted**
**1 tsp vinegar**
**½ tsp parsley flakes**
**¼ tsp dried onions, minced**
**¼ tsp garlic powder**
**¼ tsp salt**
**dash pepper**

1. Stir together milk and vinegar to substitute for buttermilk.
2. Whisk in mayonnaise until no lumps remain.
3. Whisk in remaining ingredients. Keep in tightly covered jar.

*THIS DRESSING WILL SPOIL QUICKLY WITHOUT REFRIGERATION BECAUSE IT CONTAINS MILK!*

Per serving (excluding unknown items): 69.4 Calories; 5.6g Fat (71.4% calories from fat); 0.6g Protein; 4.5g Carbohydrate; 6mg Cholesterol; 224mg Sodium.
Exchanges: ½ Fruit; 1 Fat.

# French Dressing

Makes 6 servings. Prep Time: 10 minutes

**½ c olive oil**
**3 tbsp vinegar**
**1 ½ tbsp water**
**½ tsp salt**
**¼ tsp dry mustard**
**¼ tsp paprika**

1. Shake all ingredients in a tightly covered jar. Let stand to blend flavors.
2. Shake again before serving.

**Thickened French Dressing**: Put liquids in jar. Thoroughly mix dry ingredients with 1 tbsp Instant Clearjel in a bowl. Sprinkle slowly into liquids while whisking to prevent lumping. Any of the following variations may be thickened.

**Garlic French**: Add dash freshly ground pepper and 1 clove crushed garlic ( or ¼ tsp powder). THIS DRESSING COULD DEVELOP BOTULISM IF FRESH GARLIC IS USED. USE UP QUICKLY.

**Italian Dressing**: Add ½ tsp sugar, ¼ tsp oregano, ⅛ tsp thyme, 2 cloves crushed garlic (or ½ tsp powder) and ¼ tsp onion powder ( dried onion or fresh equivalent.)

**French Tomato Dressing**: Mix ¼ c French Dressing with ¼ c catsup.

*This dressing does not contain any ingredients that would spoil quickly.*

---

Per serving (excluding unknown items): 160.7 Calories; 18.0g Fat (98.7% calories from fat); 0.0g Protein; 0.5g Carbohydrate; 0mg Cholesterol; 178mg Sodium. Exchanges: 3 ½ Fat.

## Honey-Mustard Dressing

Makes 8 servings. Prep Time: 5 minutes

**$^{1}/_{3}$ c vinegar**
**$^{2}/_{3}$ c oil**
**¼ c honey**
**2 tbsp prepared mustard**
**1 tsp salt**
**¼ tsp pepper**

1. Whisk together all ingredients until well-blended.
2. Store in a tightly covered jar. Shake before using.

*This dressing will not spoil quickly.*

Per serving (excluding unknown items): 197.2 Calories; 18.3g Fat (80.8% calories from fat); 0.2g Protein; 9.6g Carbohydrate; 0mg Cholesterol; 314mg Sodium. Exchanges: 3 ½ Fat; ½ Other Carbohydrates.

## Oriental Dressing

Makes 4 servings. Prep Time: 10 minutes

**2 tbsp vinegar**
**1 tbsp soy sauce**
**1 ½ tsp sugar**
**$^{1}/_{8}$ tsp paprika**
**½ tsp prepared mustard**
**¼ tsp salt**
**1 ½ tsp minced onion**
**¼ c oil**

1. Stir together all ingredients except oil, until sugar is dissolved.
2. Slowly add oil while beating with a whisk.

*This dressing will not spoil quickly.*

Per serving (excluding unknown items): 131.1 Calories; 13.7g Fat (91.5% calories from fat); 0.3g Protein; 2.6g Carbohydrate; 0mg Cholesterol; 398mg Sodium. Exchanges: 2 ½ Fat.

# Mayonnaise

Makes 4 servings. Prep Time: 15 minutes

**2 ½ tsp soy flour**
**1 tsp mustard**
**1 tbsp vinegar**
**pinch salt and pepper**
**½ c oil**

1. Whisk together all ingredients except oil until smooth.
2. Continue whisking vigorously, and begin adding oil, one drop at a time, then to a slow drizzle. Oil should not separate out as you are whisking. If it does, you have added to much at once and should stop adding oil and continue whisking until emulsion (oil mixed in) is reformed.
3. Mixture should have the general appearance of mayonnaise when propeı ly done. Soy flour is replacing the usual egg yolk in this recipe. Because of this, mayonnaise remains stable for only 1-2 days before oil begins to separate out.

**Egg Mayonnaise**: Omit soy flour. Substitute 2 tsp powdered eggs. This may be more stable, but egg is susceptible to spoilage and could be a source of food-borne illness. Use up quickly.

**Soyannaise**: Do not use above ingredients. Instead, combine: 1/3 c. soy milk powder (or regular powdered milk), ½ c. water, ¼ tsp salt, ¼ tsp onion powder, 1 ½ tbsp potato flakes. Whisk in 1 ½ tbsp oil. Gently stir in 2 tbsp vinegar.

*THIS DRESSING, MADE WITH SOY FLOUR, WILL NOT SPOIL QUICKLY BECAUSE IT CONTAINS NO EGG YOLK, AS DOES REGULAR MAYONNAISE.*

---

Per serving (excluding unknown items): 247.1 Calories; 27.5g Fat (98.2% calories from fat); 0.4g Protein; 0.7g Carbohydrate; 0mg Cholesterol; 16mg Sodium. Exchanges: 5 ½ Fat.

## Tomato Sauce Dressing

Makes 4 servings. Prep Time: 5 minutes

**½ c tomato sauce**
**1 tbsp vinegar**
**1 tsp dried onions**
**½ tsp Worcestershire sauce**
**¼ tsp salt**
**¼ tsp dillweed**
**¼ tsp basil**

1. Shake together all ingredients in a tightly covered jar. Shake again before serving.

*THE TOMATO SAUCE IN THIS MAY SPOIL WITHOUT REFRIGERATION.*

Per serving (excluding unknown items): 11.7 Calories; 0.1g Fat (3.3% calories from fat); 0.5g Protein; 2.9g Carbohydrate; 0mg Cholesterol; 325mg Sodium. Exchanges: ½ Vegetable.

## Vinegar and Oil Dressing

Makes 6 servings. Prep Time: 5 minutes

**¼ c olive oil**
**2 tbsp vinegar**
**½ tsp salt**
**1 clove garlic, crushed**
**dash pepper, freshly ground**

1. To use as marinade, mix all ingredients.
2. As dressing, toss about 6 c salad greens with oil until leaves glisten.
3. Mix remaining ingredients together and then toss with salad.

*IF NOT REFRIGERATED AND USING FRESH GARLIC (not garlic powder) THIS DRESSING COULD DEVELOP BOTULISM. USE UP QUICKLY. Garlic powder is safe.*

Per serving (excluding unknown items): 81.0 Calories; 9.0g Fat (97.6% calories from fat); 0.0g Protein; 0.5g Carbohydrate; 0mg Cholesterol; 178mg Sodium. Exchanges: 2 Fat.

# Enchilada Sauce

Makes 8 servings. Prep Time: 20 minutes

**4 c tomato sauce**
**4 tsp beef bouillon granules**
**½ tsp salt**
**½ tsp oregano**
**½ tsp garlic powder**
**2 tbsp chili powder, or to taste**
**½ tsp cumin**
**1 tbsp cornstarch**

1. Simmer all ingredients except cornstarch, 10-15 minutes.
2. Thicken with cornstarch which has been previously stirred with 2 tbsp water. Cook until thickened and clear.

Serve over enchiladas of all types.

Per serving (excluding unknown items): 52.0 Calories; 0.7g Fat (10.8% calories from fat); 2.2g Protein; 11.4g Carbohydrate; 0mg Cholesterol; 1191mg Sodium. Exchanges: 1 ½ Vegetable.

# Fresh Salsa

Makes 6 servings. Prep Time: 15 minutes

**2 med tomatoes, finely chopped**
**2 med jalapeno, finely chopped**
**½ c finely chopped onion**
**1 tsp salt**
**¼ tsp cumin**
**¼ tsp garlic powder**
**2 tsp chopped cilantro, if available**

1. Mix all ingredients. Serve over Taco Salad, or with chips, or any other Mexican food.

Per serving (excluding unknown items): 22.5 Calories; 0.1g Fat (4.9% calories from fat); 0.8g Protein; 5.3g Carbohydrate; 0mg Cholesterol; 361mg Sodium. Exchanges: 1 Vegetable.

# Gravy

Makes 6 servings. Prep Time: 10 minutes

**2 c beef broth**
**4 tbsp shortening**
**4 tbsp all-purpose flour**

1. Melt shortening in pan. Stir in flour. Cook and stir over low heat until mixture is smooth and bubbly.
2. Remove from heat. Stir in broth. Heat to boiling, stirring constantly. Boil and stir one minute. Sprinkle with salt and pepper to taste. Chicken broth may be used instead of beef broth.

**Milk Gravy**: Substitute milk for the beef broth. Use over biscuits, potatoes, toast, etc.
**Cornstarch Gravy**: Substitute 2 tbsp cornstarch for flour.

Serve over biscuits or vegetarian burgers.

Per serving (excluding unknown items): 114.5 Calories; 8.6g Fat (67.4% calories from fat); 4.2g Protein; 5.2g Carbohydrate; 0mg Cholesterol; 433mg Sodium. Exchanges: ½ Grain(Starch); ½ Lean Meat; 1 ½ Fat.

# Maple-flavored Syrup

Makes 6 servings. Prep Time: 10 minutes

**1 c water**
**1 ¾ c sugar**
**1 ½ tsp maple flavoring**
**1 tbsp cornstarch, optional**

1. Bring all ingredients, except cornstarch, to a boil in a 2-quart saucepan, stirring until sugar is dissolved.
2. Boil several minutes to thicken.
3. If desired, thicken also with cornstarch. Mix together 1 T cornstarch and 2 T cold water in separate dish. Stir into syrup. Stir and cook until clear.

May be used over Pancakes or Fried Mush

Per serving (excluding unknown items): 230.8 Calories; 0.0g Fat (0.0% calories from fat); 0.0g Protein; 59.5g Carbohydrate; 0mg Cholesterol; 2mg Sodium. Exchanges: 4 Fruit; 4 Other Carbohydrates.

# Hard Sauce

Makes 12 servings. Prep Time: 15 minutes

**¼ c shortening**
**1 tbsp oil**
**1 tbsp water**
**1 c powdered sugar**
**2 tsp vanilla**

1. Whip together shortening, oil and water until very fluffy, using whisk about 5 minutes.
2. Beat in 1 c powdered sugar gradually.
3. Stir in vanilla.
4. Chill as if refrigerated, if possible.

Serve over warm steamed puddings.

Per serving (excluding unknown items): 89.0 Calories; 5.4g Fat (54.5% calories from fat); 0.0g Protein; 10.2g Carbohydrate; 0mg Cholesterol; 0mg Sodium.
Exchanges: 1 Fat; ½ Other Carbohydrates.

# Vanilla Sauce

Makes 12 servings. Prep Time: 15 minutes

**¾ c sugar**
**3 tbsp cornstarch**
**¼ tsp salt**
**¼ tsp nutmeg**
**2 c water**
**1 tsp vanilla**

1. Combine sugar, cornstarch, salt and nutmeg.
2. Gradually whisk in water and cook over low heat or in double boiler until thick and clear.
3. Remove from heat and stir in vanilla.

Serve over steamed breads or puddings.

Per serving (excluding unknown items): 57.4 Calories; 0.0g Fat (0.3% calories from fat); 0.0g Protein; 14.5g Carbohydrate; 0mg Cholesterol; 46mg Sodium.
Exchanges: 1 Fruit; 1 Other Carbohydrates.

## Seasoned Tomato Sauce

Makes 6 servings. Prep Time: 30 minutes

**¼ tsp garlic powder**
**½ c finely chopped onion**
**1 tbsp olive oil**
**2 pt canned tomatoes**
**1 pt tomato sauce**
**1 tsp sugar**
**½ tsp basil**
**½ tsp rosemary**
**salt to taste**

1. Saute onion and garlic in oil in a 2-quart saucepan over medium heat.
2. Add tomatoes and sauce. Stir in seasonings.
3. Bring to a boil. Reduce heat, cover and simmer at least 20 minutes, or up to 2 hours. Stir occasionally. Flavor improves with cooking.

**Marinara Sauce**: Omit onion, sugar, basil and rosemary. Add 1 tsp oregano and 1 tbsp parsley flakes.

---

Per serving (excluding unknown items): 28.1 Calories; 1.0g Fat (28.3% calories from fat); 0.8g Protein; 4.9g Carbohydrate; 0mg Cholesterol; 7mg Sodium.
Exchanges: 1 Vegetable.

## Catsup

Makes 20 servings. Prep Time: 1 hour

**4 c tomato sauce**
**¼ tsp cinnamon**
**¼ tsp cloves**
**¼ tsp dry mustard**
**½ tsp celery seed**
**½ c vinegar**
**½ c chopped onion**
**⅛ tsp cayenne pepper**
**½ c sugar**
**1 ½ tsp salt**

1. Combine all except vinegar and salt in a 2 quart saucepan.
2. Simmer until reduced by almost half. May be sieved to remove onion.
3. Add vinegar and salt.
4. Simmer to desired consistency. Allow to cool.

Makes about 2 ½ cups, 2 tablespoon servings.

---

Per serving (excluding unknown items): 36.7 Calories; 0.1g Fat (2.4% calories from fat); 0.7g Protein; 9.3g Carbohydrate; 0mg Cholesterol; 457mg Sodium.
Exchanges: ½ Vegetable; ½ Other Carbohydrate.

# White Sauce, Medium

Makes 6 servings. Prep Time: 15 minutes

---

**1 tbsp shortening**
**2 tbsp all-purpose flour**
**¼ tsp salt**
**⅛ tsp pepper**
**1 c milk, reconstituted**

1. Heat shortening in saucepan over low heat until melted.
2. Blend in flour, salt and pepper until smooth. Cook, stirring constantly until bubbly. Remove from heat.
3. Whisk in milk slowly. Heat to boiling, stirring constantly. Boil and stir one minute, until thick.

**Cornstarch White Sauce**: Substitute 1 T cornstarch for all-purpose flour.

**Thin White Sauce**: Use 1 T flour or 1 ½ tsp cornstarch.

**Thick White Sauce**: Use 4T flour or 2T cornstarch.

Serve over vegetables or pasta.

---

Per serving (excluding unknown items): 53.5 Calories; 3.5g Fat (59.0% calories from fat); 1.6g Protein; 3.9g Carbohydrate; 6mg Cholesterol; 109mg Sodium.
Exchanges: ½ Fat.

# SOUPS, STEWS AND CHILI

Soups, stews and chili were the mainstay of pre-industrial America. Typically, a pot of soup was started over the fire in the fireplace in the fall, and the pot was never empty until spring. On a daily basis, ingredients would be added, new combinations tried, and meals removed from the pot. As long as the fire was going, the food was cooking. Was this the first fast food?

Many of these recipes include lentils and split peas because they are ideal legumes for low cost and low fuel requirements. Other recipes use beans or TVP. Six different chili recipes are provided. "Chicken" Chili is the most traditional, with the exception of its use of chicken-flavored TVP instead of ground beef. Texas Chili uses no tomatoes, only chili powder to achieve its red color. The remaining recipes use lentils or whole wheat in place of some or all of the usual beans. If you like your chili extra hot, try Four-Alarm Chili. It is sure to warm you up!

A good addition to any soup, stew or chili is dumplings. These are like a biscuit, but they are cooked right on top of the boiling stew. Look for this recipe in the Stovetop Breads chapter. Toasted Croutons or Basic Popcorn are also good replacements for the usual soda crackers on top of soup.

To save on fuel requirements, consider using an insulated box to complete the cooking of these dishes. Start cooking your soup, stew or chili in a three to six-quart pot with a tight-fitting lid. Bring it to a boil and cook about 15 minutes. Then put it in the hot box. Arrange the insulating material snugly around it, avoiding air pockets. Place a pillow over the top and go relax. Let your dinner cook for three to four hours. For the hot box, use a container, such as a cardboard box or laundry basket, that is large enough to allow you to fit at least four inches of insulating material on all sides of your pot. Insulation could be shredded newspaper, straw, sawdust, dried lawn clippings, an old quilt or an old cotton-covered sleeping bag (nylon could melt.) Any material that could be messy may be placed in cloth bags or pillowcases. To keep the pot as hot as possible prior to the hot box, be sure not to lift the lid of the pot for 5 or 10 minutes prior to removing it from the stove. After removal, place it immediately in the hot box and cover it.

# "Beef" and Barley Soup

Makes 10 servings. Prep Time: 1 hour

**2 qt water**
**1 c pearl barley**
**1 c textured soy protein, cubed**
**3 lg carrots, diced**
**1 lg onion, chopped**
**1 c Swiss chard, chopped stalks**
**2 tbsp beef bouillon granules**
**1 tsp salt**
**¼ tsp pepper**
**1 each bay leaf**
**½ tsp marjoram, optional**
**½ tsp thyme, optional**
**1 qt canned tomatoes**

1. Combine all ingredients in a 6-quart pot. If unflavored TVP is used, additional beef bouillon granules may be needed for flavor.
2. Bring to a boil. Reduce heat and simmer, about 50 minutes, until tender.

Per serving (excluding unknown items): 181.7 Calories; 0.9g Fat (4.2% calories from fat); 19.1g Protein; 28.8g Carbohydrate; 0mg Cholesterol; 800mg Sodium. Exchanges: 1 ½ Grain(Starch); 2 Lean Meat; 1 ½ Vegetable.

# Almost-Chicken Soup with Rice

Makes 8 servings. Prep Time: 1 hour

**2 qt water**
**2 tbsp chicken bouillon granules**
**1 c textured soy protein, chunks**
**1 c chopped onion**
**3 med carrots, julienned**
**1 pt peas, canned**
**1 pt green beans, canned**
**1 c rice**
**parsley, to taste**
**salt and pepper, to taste**

1. In a 6-quart pot, bring water to boil and add granules, soy protein, onion and carrots. Cover and simmer until vegetables are tender.
2. About 30 minutes before serving, add rice and simmer. Add seasonings.
3. Cook 15-20 minutes, until rice is tender. Add canned vegetables last.

Per serving (excluding unknown items): 236.6 Calories; 1.0g Fat (3.6% calories from fat); 24.3g Protein; 36.7g Carbohydrate; 0mg Cholesterol; 513mg Sodium. Exchanges: 2 Grain(Starch); 2 ½ Lean Meat; 1 Vegetable.

# Classic Split Pea Soup

Makes 6 servings. Prep Time: 1 hour and 15 minutes

**2 c split peas, green or yellow**
**8 c chicken broth, from base**
**1 c chopped onion**
**1 c diced carrot**
**2 med potatoes, peeled and diced**
**2 each bay leaves**
**1 c soy bacon bits, reconstituted**
**salt and pepper, to taste**

1. Combine all ingredients in a large pot, 4 to 6 quart. Bring to a boil.
2. Cover and simmer, stirring occasionally, until vegetables are tender and soup thickens, about 1 hour.
3. Remove bay leaves. Adjust salt and pepper to taste.

Per serving (excluding unknown items): 616.3 Calories; 14.7g Fat (20.8% calories from fat); 46.5g Protein; 79.5g Carbohydrate; 3mg Cholesterol; 2809mg Sodium. Exchanges: 5 Grain(Starch); 4 ½ Lean Meat; ½ Vegetable; 1 Fat.

# Corn Chowder

Makes 6 servings. Prep Time: 25 minutes

**2 tbsp oil**
**½ c soy bacon bits, soaked**
**1 c chopped onion**
**½ c Swiss chard, chopped stalks**
**2 med chopped carrots**
**2 c potatoes, finely chopped**
**1 pt canned corn**
**3 c milk, reconstituted**
**2 tbsp cornstarch**
**1 ¼ tsp salt**
**⅛ tsp pepper**

1. Cook onion, chard, carrots and potatoes in 2 tbsp oil in a 3-quart pot, about 8 minutes until tender-crisp.
2. Add remaining ingredients except cornstarch and heat almost to a boil.
3. Meanwhile, stir together cornstarch and about ¼ c water in a separate bowl, until smooth.
4. Pout into hot soup and continue to heat, stirring, until soup is thickened.

Per serving (excluding unknown items): 351.7 Calories; 14.4g Fat (34.8% calories from fat); 14.3g Protein; 46.2g Carbohydrate; 17mg Cholesterol; 1056mg Sodium. Exchanges: 2 ½ Grain(Starch); ½ Lean Meat; ½ Non-Fat Milk; 1 Vegetable; 2 ½ Fat.

# Cowpuncher Stew

Makes 6 servings. Prep Time: 45 minutes

**4 c pinto beans, cooked**
**2 tbsp oil**
**1 c chopped onion**
**½ tsp garlic powder**
**1 pt canned tomatoes**
**½ c green chiles, chopped**
**5 c water**
**3 tsp beef bouillon granules**
**½ tsp cumin**
**2 lg potatoes, peeled and chopped**
**1 pt green beans, canned**
**1 pt canned corn**

1. Cook onion in hot oil until tender in a 4-quart pot
2. Add remaining ingredients and simmer 30-45 minutes, until tender.

Per serving (excluding unknown items): 381.9 Calories; 6.2g Fat (13.8% calories from fat); 15.1g Protein; 71.4g Carbohydrate; 0mg Cholesterol; 667mg Sodium. Exchanges: 4 Grain(Starch); ½ Lean Meat; 1 Vegetable; 1 Fat.

# Hearty Lentil Stew

Makes 8 servings. Prep Time: 1 hour

**2 tbsp oil**
**1 c chopped onion**
**7 c water**
**2 tbsp chicken bouillon granules**
**1 c lentils, rinsed**
**½ c elbow macaroni**
**2 each carrots, sliced**
**2 c garbanzo beans, cooked**
**½ tsp thyme, leaves removed**

1. Cook onion in hot oil in a 4 or 6 quart pot over medium-high heat, stirring often, until well browned.
2. Add water, bouillon granules and lentils. Cover and simmer 30 minutes.
3. Add remaining ingredients, cover and simmer 15-20 minutes more, until carrots and lentils are tender.

Per serving (excluding unknown items): 214.3 Calories; 5.2g Fat (21.3% calories from fat); 11.7g Protein; 31.7g Carbohydrate; 0mg Cholesterol; 507mg Sodium. Exchanges: 2 Grain(Starch); ½ Lean Meat; ½ Vegetable; 1 Fat.

# Italian Garden Pea Soup

Makes 8 servings. Prep Time: 2 hours

**1 ½ c split peas, green or yellow**
**½ c navy beans, soaked**
**3 c tomato juice**
**½ c chopped onion**
**1 c zucchini, cubed**
**2 c cabbage, coarsely chopped**
**1 each turnip, diced, optional**
**1 c diced carrot**
**½ tsp garlic powder**
**1 tsp salt, or to taste**
**½ tsp pepper**
**½ tsp oregano**
**12 tsp basil**
**4 oz spaghetti, uncooked**

1. Combine split peas, navy beans and 8 c water in a 6 quart pot. Bring to a boil. Reduce heat, cover and simmer 1 to 1 ½ hours, until beans are tender.
2. Add remaining ingredients except spaghetti. Cook until vegetables are tender, 20-30 minutes.
3. Break spaghetti in quarters. Add to soup and cook 8-10 minutes more.

Per serving (excluding unknown items): 265.1 Calories; 1.1g Fat (3.6% calories from fat); 15.7g Protein; 50.9g Carbohydrate; 0mg Cholesterol; 626mg Sodium. Exchanges: 3 Grain(Starch); 1 Lean Meat; 1 ½ Vegetable.

## Lentil and Greens Soup

Makes 6 servings. Prep Time: 1 hour

**2 ½ c lentils, rinsed**
**2 tbsp oil**
**1 lg onion, finely chopped**
**salt and pepper, to taste**
**1 each bay leaf**
**water**
**chicken bouillon granules, optional**
**4 c fresh spinach, chopped**

1. In a 3-quart pot, heat oil and cook onion until tender, about 5 minutes.
2. Add remaining ingredients except spinach. Add water to cover by about 2 inches. Add bouillon, if desired.
3. Simmer until lentils are soft, about 40 minutes.
4. Remove bay leaf. Add spinach and cook 10 more minutes.

*Swiss chard or other greens may be substituted for the spinach leaves.*

Per serving (excluding unknown items): 329.1 Calories; 5.5g Fat (14.5% calories from fat); 23.8g Protein; 49.3g Carbohydrate; 0mg Cholesterol; 38mg Sodium. Exchanges: 3 Grain(Starch); 2 Lean Meat; ½ Vegetable; 1 Fat.

## Onion-Lentil Soup

Makes 6 servings. Prep Time: 50 minutes

**4 c sweet onions, thinly sliced**
**½ tsp garlic powder**
**3 tbsp oil**
**1 c lentils, rinsed**
**2 c carrots, coarsely grated**
**1 tsp thyme**
**1 each bay leaf**
**4 c beef broth, from base**
**salt, to taste**
**pepper, to taste**
**1 c bread crumbs**
**3 tbsp shortening**

1. In a 6-quart pot, saute onion in oil until golden, but not browned.
2. Add all ingredients except bread crumbs and remaining shortening. Simmer 45 minutes, stirring occasionally. Remove bay leaf.
3. Brown bread crumbs in shortening until golden. Sprinkle atop soup.

Per serving (excluding unknown items): 390.5 Calories; 15.0g Fat (33.6% calories from fat); 20.8g Protein; 46.0g Carbohydrate; 0mg Cholesterol; 1041mg Sodium. Exchanges: 2 Grain(Starch); 1 ½ Lean Meat; 2 Vegetable; 3 Fat.

# Lentil Carrot Soup

Makes 6 servings. Prep Time: 1 hour

**4 c carrots, coarsely grated**
**1 ½ c chopped onion**
**½ tsp garlic powder**
**2 tbsp oil**
**2 ½ c lentils, rinsed**
**3 qt water**
**½ tsp thyme**
**salt, to taste**
**1/8 tsp cayenne pepper**
**1 c croutons, or popcorn**

1. Saute carrots and onions in oil until soft, about 5 minutes.
2. Stir in lentils and saute 1-2 minutes.
3. Add water, bring to a boil and simmer, uncovered, over medium heat until lentils are very soft and soup is thickened, about 45 minutes.
4. Stir in spices and heat through.

Per serving (excluding unknown items): 378.8 Calories; 5.9g Fat (13.5% calories from fat); 24.3g Protein; 60.5g Carbohydrate; 0mg Cholesterol; 84mg Sodium. Exchanges: 3 ½ Grain(Starch); 2 Lean Meat; 2 Vegetable; 1 Fat.

# Lentil Barley Soup

Makes 8 servings. Prep Time: 1 hour

**1 c chopped onion**
**¼ tsp garlic powder**
**2 tbsp oil**
**1 qt canned tomatoes, or diced fresh**
**¾ c lentils, rinsed**
**¾ c pearl barley**
**6 c water**
**6 tsp beef broth cubes**
**½ tsp rosemary**
**½ tsp oregano**
**¼ tsp pepper**
**2 c carrots, thinly sliced**

1. In a 4 or 6 quart soup pot, cook onion in hot oil until tender.
2. Add remaining ingredients and cook at least 40 minutes, until the barley, lentils and carrots are tender.

Per serving (excluding unknown items): 207.6 Calories; 4.3g Fat (18.0% calories from fat); 9.2g Protein; 35.3g Carbohydrate; 0mg Cholesterol; 1162mg Sodium. Exchanges: 1 ½ Grain(Starch); ½ Lean Meat; 1 ½ Vegetable; ½ Fat.

# New Year's Eve Lentil Soup

Makes 10 servings. Prep Time: 2 hours

**¼ c soy bacon bits, soaked**
**¾ c diced onion**
**¾ c diced carrot**
**¾ c Swiss chard, diced stalks**
**¾ c flour**
**3 ½ qt water**
**¾ c potatoes, diced**
**4 tbsp oil**
**1 c lentils, rinsed**
**2 tsp salt**
**2 each bay leaves**
**1 tsp thyme**
**3 tbsp beef bouillon granules**
**pinch nutmeg**
**pinch pepper**

1. In a 6-quart pot, saute onion, chard and carrots in oil until soft.
2. Add flour, stirring constantly to blend smoothly. All-purpose or soft white flour works best. $^{1}/_{3}$ c cornstarch could also be substituted.
3. Slowly add the water, stirring constantly. Then add remaining ingredients.
4. Simmer for 2-3 hours. Flavor improves with time. Remove bay leaves. Ham TVP may be substituted for soy bacon bits.

*Eating lentils on New Year's Eve to assure prosperity and good fortune is a tradition of the Greeks and Italians.*

---

Per serving (excluding unknown items): 205.3 Calories; 7.7g Fat (32.6% calories from fat); 9.4g Protein; 26.1g Carbohydrate; 0mg Cholesterol; 1088mg Sodium. Exchanges: 1 ½ Grain(Starch); ½ Lean Meat; ½ Vegetable; 1 ½ Fat.

# Palouse Paradise Soup

Makes 6 servings. Prep Time: 1 hour

**1 c split peas, rinsed**
**1 c lentils, rinsed**
**4 c water**
**1 c soy bacon bits, soaked**
**1 c chopped carrots**
**1 med onion, finely chopped**
**½ tsp garlic powder**
**3 c milk, reconstituted**
**6 sprigs parsley, chopped**
**4 tbsp potato flakes, optional**
**2 tsp salt**
**½ tsp pepper**
**½ tsp coriander**
**½ tsp cumin**
**¼ tsp turmeric**
**¼ tsp marjoram**

1. Cook peas and lentils in water in a 6-quart pot for about 20 minutes.
2. Add remaining ingredients except milk. Cover and simmer for about 30 minutes, or until vegetables are tender.
3. Add milk just before serving.

*The Palouse area of eastern Washington produces about 95% of all the lentils and split peas grown in the US.*

---

Per serving (excluding unknown items): 553.5 Calories; 16.1g Fat (24.6% calories from fat); 39.1g Protein; 71.7g Carbohydrate; 17mg Cholesterol; 1585mg Sodium. Exchanges: 3 ½ Grain(Starch); 3 Lean Meat; ½ Non-Fat Milk; 3 Vegetable; 2 Fat.

# Potato Soup

Makes 6 servings. Prep Time: 1 hour and 10 minutes

**6 med potatoes, peeled and sliced**
**2 med onions, finely chopped**
**3 c water**
**2 c milk, double-strength**
**1 tbsp oil**
**salt and pepper, to taste**
**1 tbsp chives or parsley, chopped**
**2 c broccoli, or other vegetables**

1. Cook potatoes and onions in 3 c water, covered, for 45 minutes. Mash.
2. Stir in remaining ingredients. Simmer 15 minutes. Do not boil.

**"Bacon" Potato Soup**: Add 4 tsp ham or chicken broth base and ¼ c ham or bacon flavored TVP when adding milk.
**Instant Potato Soup**: Use potato flakes to make or thicken soup as desired.

Per serving (excluding unknown items): 371.8 Calories; 5.4g Fat (12.8% calories from fat); 10.7g Protein; 72.4g Carbohydrate; 11mg Cholesterol; 66mg Sodium. Exchanges: 4 ½ Grain(Starch); 1 Vegetable; 1 Fat.

# Tuna Chowder

Makes 6 servings. Prep Time: 30 minutes

**4 tbsp oil**
**½ c Swiss chard, chopped stalks**
**1 ½ c chopped onion**
**2 c potato, diced**
**2 tsp salt**
**¼ tsp pepper**
**2 tbsp parsley**
**½ tsp thyme**
**½ tsp dill weed**
**4 tbsp cornstarch**
**1 c canned tomatoes**
**5 c milk, reconstituted**
**1 can tuna in water, drained**

1. In a 6-quart pot, cook chard, onion and potato in oil over medium heat about 15 minutes, until tender.
2. Stir in remaining ingredients except cornstarch. Heat,stirring until soup almost comes to a boil.
3. Meanwhile, mix cornstarch and ½ c water in a separate bowl, til smooth.
4. Pour into hot soup and continue to heat, stirring, until soup is thickened.

Per serving (excluding unknown items): 341.1 Calories; 16.3g Fat (42.5% calories from fat); 15.2g Protein; 34.4g Carbohydrate; 35mg Cholesterol; 988mg Sodium. Exchanges: 1 ½ Grain(Starch); 1 Lean Meat; ½ Non-Fat Milk; 1 Vegetable; 3 Fat.

# Tuscany Bean Soup

Makes 6 servings. Prep Time: 30 minutes

**2 c white beans, cooked**
**1 c chopped spinach**
**1 lg carrot, sliced**
**1 med onion, sliced**
**¾ tsp garlic powder**
**6 c chicken broth, from base**
**1 tsp basil**
**¼ tsp pepper**
**1 ½ tsp parsley**
**2 tbsp oil**

1. In a 4-quart pot, heat oil over medium heat. Saute onion and carrot until tender, about 5 minutes.
2. Add remaining ingredients and simmer until heated through and slightly thickened, about 20 minutes.

---

Per serving (excluding unknown items): 360.7 Calories; 7.8g Fat (19.1% calories from fat); 27.6g Protein; 46.6g Carbohydrate; 3mg Cholesterol; 1589mg Sodium. Exchanges: 2 ½ Grain(Starch); 2 ½ Lean Meat; ½ Vegetable; 1 Fat.

# "Chicken" Chili

Makes 6 servings. Prep Time: 45 minutes

**2 tbsp oil**
**½ c chopped onion**
**½ c chopped green bell pepper**
**¼ tsp garlic powder**
**2 c cooked kidney beans**
**1 pt canned tomatoes**
**1 c tomato sauce**
**¼ tsp dry mustard**
**1 c canned corn**
**2 tbsp vinegar**
**2 tsp Worcestershire sauce**
**½ c textured soy protein, chunks**
**1 tbsp brown sugar**
**1 tsp chili powder**

1. In a 2 or 3-quart pot, cook onion and green pepper in hot oil until tender-crisp.
2. Add remaining ingredients, stirring to break up tomatoes.
3. Cover and simmer for 25-30 minutes, until flavors are blended and chili is hot.

---

Per serving (excluding unknown items): 249.3 Calories; 5.6g Fat (18.5% calories from fat); 20.4g Protein; 35.3g Carbohydrate; 0mg Cholesterol; 534mg Sodium. Exchanges: 1 ½ Grain(Starch); 2 Lean Meat; 1 ½ Vegetable; 1 Fat.

## Easy Lentil Chili

Makes 6 servings. Prep Time: 1 hour

**2 ½ c lentils, rinsed**
**5 c water**
**1 pt canned tomatoes**
**2 tbsp onion flakes**
**4 tsp beef bouillon granules**
**1 ½ tsp chili powder**
**½ tsp cumin**

1. Simmer lentils and water in a 3 or 4-quart saucepan for 30 minutes. Do not drain.
2. Add remaining ingredients and simmer 30 minutes more.

May be served over cornbread or baked potatoes.

Per serving (excluding unknown items): 299.9 Calories; 1.3g Fat (3.8% calories from fat); 23.8g Protein; 51.5g Carbohydrate; 0mg Cholesterol; 588mg Sodium. Exchanges: 3 Grain(Starch); 2 Lean Meat; ½ Vegetable.

## Lentil Chili Deluxe

Makes 10 servings. Prep Time: 50 minutes

**5 c water**
**2 c garbanzo beans, cooked**
**1 c chopped onion**
**½ tsp garlic powder**
**2 tsp cumin**
**1 tsp salt**
**2 ½ c lentils, rinsed**
**2 c kidney beans, cooked**
**1 pt canned tomatoes, or diced fresh**
**½ c chopped carrots**
**½ c chopped green bell pepper**
**1 tbsp chili powder**
**1 tsp cayenne pepper**

1. In a 6-quart, heavy pot, combine all ingredients.
2. Cover and bring to a boil.
3. Simmer 30-45 minutes, until lentils are tender.

Per serving (excluding unknown items): 285.3 Calories; 1.9g Fat (5.8% calories from fat); 20.4g Protein; 49.6g Carbohydrate; 0mg Cholesterol; 338mg Sodium. Exchanges: 3 Grain(Starch); 1 ½ Lean Meat; ½ Vegetable; ½ Fat.

# Four-alarm Chili

Makes 10 servings. Prep Time: 2 hours

**1 each red bell pepper, finely chopped**
**1 each green bell pepper, finely chopped**
**1 med onion, finely chopped**
**1 med carrot, finely chopped**
**1 med jalapeno, finely chopped**
**3 cloves garlic, finely chopped**
**4 tbsp olive oil, for frying**
**7 c water**
**1 lb lentils, rinsed**
**¾ c tomato sauce**
**2 c pinto beans, cooked**
**2 c kidney beans, cooked**
**⅓ c chili powder, or to taste**
**4 tsp cumin**
**¼ tsp red pepper flakes**
**salt and pepper, to taste**
**1 qt canned tomatoes**

1. Saute finely chopped vegetables in olive oil in heavy 6-quart pot on low heat until softened.
2. Add water and lentils. Simmer 45 minutes, until lentils are tender.
3. Add remaining ingredients and simmer 45 minutes.

*This is a favorite recipe of firemen from a particular firehouse in New York City. The "four alarm" refer to the four sources of "heat" in this HOT chili.*

Per serving (excluding unknown items): 349.3 Calories; 7.4g Fat (18.0% calories from fat); 20.9g Protein; 54.5g Carbohydrate; 0mg Cholesterol; 372mg Sodium. Exchanges: 3 Grain(Starch); 1 ½ Lean Meat; 1 ½ Vegetable; 1 ½ Fat.

# Quick Whole Wheat Chili

Makes 6 servings. Prep Time: 40 minutes

**6 c whole wheat berries, cooked**
**1 c chopped onion**
**½ tsp garlic powder**
**2 tbsp oil**
**1 ½ tsp chili powder**
**1 tsp cumin**
**1 tsp salt**
**1 pt canned tomatoes**
**1 tbsp beef bouillon granules**
**1 c tomato sauce, optional**

1. Saute onion in hot oil in a 3 or 4-quart pot until tender.
2. Add remaining ingredients and simmer for about 30 minutes, to blend flavors. Add tomato sauce, up to 2 c, if needed to "thin" the chili.

Per serving (excluding unknown items): 237.7 Calories; 5.6g Fat (19.8% calories from fat); 7.7g Protein; 43.5g Carbohydrate; 0mg Cholesterol; 1087mg Sodium. Exchanges: 2 ½ Grain(Starch); 1 ½ Vegetable; 1 Fat.

# Texas Chili

Makes 6 servings. Prep Time: 40 minutes

**6 c kidney beans, cooked**
**2 tbsp oil**
**4 tbsp chili powder, to taste**
**2 tsp cumin**
**3 tbsp cornstarch**
**1 c chopped onion**
**3 c beef broth, from base**
**½ tsp oregano**
**½ tsp garlic powder**
**1 tsp salt**

1. Cook onion in hot oil in a 3-quart saucepan until tender.
2. Add remaining ingredients. Cover and simmer 30 minutes.

*This chili gets its red color from the chili powder, not from tomatoes!*

Per serving (excluding unknown items): 339.7 Calories; 6.5g Fat (16.6% calories from fat); 21.9g Protein; 51.4g Carbohydrate; 0mg Cholesterol; 1062mg Sodium. Exchanges: 3 Grain(Starch); 1 ½ Lean Meat; ½ Vegetable; 1 Fat.

# STOVETOP MAIN DISHES

This chapter is heavy on the Italian and Mexican menus, with a good bit of Chinese food as well. Some require two pots, one for the pasta and one for the rest of the dish. Others take only one pot, so, depending on your fuel and clean up needs, choose accordingly.

Four recipes for vegetarian burgers or patties have been included. These are nice alternatives for those who are still experiencing withdrawal from burgers and fries, but they are also a good way to use up leftover foods. Whole Wheat Patties, in particular, can use up just about anything you have: beans, vegetables, fish or TVP. Homemade hamburger buns (see Basic Whole Wheat Bread: Theme and Variations) and catsup make the treat complete.

# Bean Stroganoff

Makes 6 servings. Prep Time: 45 minutes

---

**2 med onions, sliced**
**1 tbsp oil**
**2 tbsp cornstarch**
**1 c beef broth, from base**
**4 tsp Worcestershire sauce**
**1/8 tsp chili powder**
**1/8 tsp marjoram**
**1/8 tsp thyme**
**dash nutmeg**
**½ tsp garlic powder**
**3 c pinto beans, cooked**
**1 c evaporated skim milk (see note)**

1. Saute onions in oil in a large skillet until tender.
2. Mix cornstarch (or ¼ c all-purpose flour) in a small amount of the broth until smooth; then add remaining broth, Worcestershire and seasonings.
3. Add to skillet and cook until thick. Stir in beans and stir over low heat until heated through.
4. Remove beans from heat. Stir in evaporated milk. Serve over hot cooked cracked wheat, rice or noodles.

NOTE: Prepare evaporated skim milk from powdered milk by combining 1 scant cup water with 6 tbsp powdered milk.

*Evaporated skim milk replaces sour cream in this recipe.*

---

Per serving (excluding unknown items): 212.6 Calories; 2.9g Fat (12.0% calories from fat); 12.7g Protein; 34.8g Carbohydrate; 2mg Cholesterol; 302mg Sodium. Exchanges: 1 ½ Grain(Starch); ½ Lean Meat; ½ Non-Fat Milk; ½ Vegetable; ½ Fat.

# Browned Rice

Makes 4 servings. Prep Time: 45 minutes

**1 c rice**
**¼ c shortening**
**½ c chopped onion**
**1 tsp salt**
**3 ½ c water**

1. Heat shortening in large skillet. Add rice. Cook and stir constantly until rice is lightly browned, about 10 minutes.
2. Add onions (and/or other vegetables, as desired) and cook 2-3 minutes.
3. Add salt and water. Bring to a boil. Reduce heat, cover and simmer 20-25 minutes until rice is tender and liquid is absorbed.

Serve with beans or milk for a complete meal.

Per serving (excluding unknown items): 289.7 Calories; 13.1g Fat (41.2% calories from fat); 3.5g Protein; 38.7g Carbohydrate; 0mg Cholesterol; 542mg Sodium. Exchanges: 2 ½ Grain(Starch); 2 ½ Fat.

# Garlic Rice and Pasta

Makes 6 servings. Prep Time: 40 minutes

**2 tbsp shortening**
**1 c rice**
**½ c spaghetti, broken up**
**3 c water**
**2 tbsp sliced green onions**
**½ tsp garlic powder**
**½ tsp salt**
**1/8 tsp pepper**
**1 tbsp parsley flakes**

1. In a large skillet, heat shortening until hot. Add rice and spaghetti. Cook and stir until spaghetti is golden brown.
2. Add remaining ingredients. Stir. Reduce heat, cover and simmer about 30 minutes until rice is tender and liquid is absorbed.

Serve with legumes or milk for protein.

Per serving (excluding unknown items): 178.1 Calories; 4.6g Fat (23.5% calories from fat); 3.2g Protein; 30.3g Carbohydrate; 0mg Cholesterol; 184mg Sodium. Exchanges: 2 Grain(Starch); 1 Fat.

# East Indian Cheese Curry

Makes 6 servings. Prep Time: 50 minutes

**1 ¼ c Pressed Easy Cheese**
**2 tbsp oil**
**1 lg onion, chopped**
**1 tsp garlic powder**
**1 tsp ginger**
**2 med tomatoes**
**½ tsp cumin**
**1 tsp ground coriander**
**1 tsp turmeric**
**¼ tsp cayenne, or to taste**
**¼ tsp fennel seed, crushed**
**1 ½ c whey**
**1 tbsp cilantro**
**salt, to taste**
**6 c cooked rice**
**½ c chopped nuts, optional**

1. Cut cheese into ½-inch cubes. Set aside.
2. In a large skillet, heat oil over medium-high heat, and cook onion, garlic powder and ginger, stirring, until lightly brown, about 10 minutes.
2. Add tomatoes (fresh or canned), cumin, coriander, turmeric, cayenne and fennel. Cook and stir until tomato is very soft and does not hold its shape, about 10 minutes.
3. Stir in whey (or water) and boil, stirring until sauce looks completely mixed, 3-5 minutes. Stir in cilantro and salt.
4. Cover, reduce heat to low, and simmer 20 minutes.
5. Gently stir in cheese and cook, covered, about 3 minutes to warm cheese.
6. Serve over rice and top with nuts.

*This dish contains traditional Indian spices. If these are unavailable, or if they don't suit your taste, try the dish without any extra spices and see how you like it—then try adding some of your own.*

Per serving (excluding unknown items): 516.4 Calories; 12.5g Fat (21.7% calories from fat); 17.1g Protein; 84.4g Carbohydrate; 4mg Cholesterol; 339mg Sodium. Exchanges: 3 ½ Grain(Starch); 1 ½ Lean Meat; 1 Vegetable; 2 Fat; 1 ½ Other Carbohydrates.

# Fried Cheese

Makes 4 servings. Prep Time: 20 minutes

**1 ¼ c Pressed Easy Cheese**
**1 med egg, reconstituted**
**¼ c fine dry breadcrumbs**
**2 tbsp oil**

1. Divide and shape cheese into 16 small triangles ¼-inch thick.
2. Dip into egg, then into breadcrumbs.
3. Heat oil in a 10 or 12-inch skillet on medium heat. Add cheese and brown.

---

Per serving (excluding unknown items): 143.8 Calories; 8.6g Fat (54.6% calories from fat); 10.2g Protein; 5.9g Carbohydrate; 56mg Cholesterol; 80mg Sodium. Exchanges: ½ Grain(Starch); 1 ½ Lean Meat; 1 ½ Fat.

# Fried Rice

Makes 6 servings. Prep Time: 20 minutes

**1 med onion, chopped**
**¼ c chopped green bell pepper**
**4 tbsp oil**
**3 c cooked rice**
**½ c Swiss chard, chopped stalks**
**½ c grated carrots**
**3 tbsp soy sauce**

1. In a 10-inch skillet, cook and stir onion, green pepper and chard until tender.
2. Stir in rice, carrot and soy sauce. Cook over low heat, stirring frequently, 5-7 minutes.

**Fried Wheat**: Substitute 3 cups cooked whole wheat berries for the rice.

*Add cooked legumes, TVP or chunks of cheese for complementary proteins; or serve with a glass of milk.*

---

Per serving (excluding unknown items): 222.6 Calories; 9.4g Fat (38.3% calories from fat); 3.5g Protein; 30.7g Carbohydrate; 0mg Cholesterol; 527mg Sodium. Exchanges: 1 ½ Grain(Starch); 1 Vegetable; 2 Fat.

# Grand Prize Zucchini Skillet

Makes 6 servings. Prep Time: 45 minutes

**1 c lentils, rinsed**
**2 c water**
**1 med onion, chopped**
**2 tbsp oil**
**2 med zucchini, sliced**
**2 med tomatoes, sliced**
**½ tsp garlic powder**
**½ tsp salt**
**¼ tsp pepper**
**1 tbsp soy sauce**

1. In large skillet, cook lentils in water for 30 minutes; drain and remove lentils from pan temporarily.
2. Saute onion in oil until it is tender-crisp. Add lentils and zucchini and saute 5 minutes more.
3. Stir in seasonings. Arrange tomatoes on top and steam, covered, until tomatoes are tender, 5-10 minutes.

---

Per serving (excluding unknown items): 177.0 Calories; 5.0g Fat (24.4% calories from fat); 10.3g Protein; 24.7g Carbohydrate; 0mg Cholesterol; 360mg Sodium. Exchanges: 1 Grain(Starch); ½ Lean Meat; 1 Vegetable; 1 Fat.

# Hot Mexican Salad

Makes 4 servings. Prep Time: 25 minutes

**1 can canned corn**
**¼ c chopped onion**
**2 c cooked red kidney beans**
**½ c Catalina salad dressing**
**1 tbsp chili powder**
**4 c shredded lettuce**
**½ c sliced green onions**
**9 oz tortilla chips**

1. Saute onion in 2 tbsp oil or shortening until tender.
2. Stir in corn (undrained), beans, dressing, chili powder; simmer 15 mins.
3. Combine lettuce and green onion. Pour bean mixture on top and toss lightly. Serve with tortilla chips.

*Catalina French Dressing is recommended for flavor in this recipe.*

---

Per serving (excluding unknown items): 610.5 Calories; 30.7g Fat (43.5% calories from fat); 34.6g Protein; 75.5g Carbohydrate; 4mg Cholesterol; 861mg Sodium. Exchanges: 4 ½ Grain(Starch); ½ Lean Meat; ½ Vegetable; ½ Fruit; 6 Fat.

# Lentil Burgers

Makes 6 servings. Prep Time: 45 minutes

**1 c lentils, rinsed**
**2 c water**
**1 ½ c soft bread crumbs**
**½ c finely chopped onion**
**1 tsp salt**
**2 lg eggs, reconstituted**
**dash Tabasco sauce**
**2 tsp beef bouillon granules**
**6 tbsp oil, for frying**

1. Cook lentils in water until soft, about 40 minutes. Drain. Mash slightly.
2. Add remaining ingredients and form patties using ½ c mixture each.
3. Fry in oil or shortening in a large skillet until brown on both sides, about 5 minutes per side.

NOTE: An easy way to make patties is to use two straight-sided, flat-bottomed bowls that telescope into each other. Sprinkle some bread crumbs or cornmeal into one bowl to cover the bottom (keeps patties from sticking.) Place ½ c patties mixture into bowl and flatten slightly with spoon. Sprinkle patties with bread crumbs to cover. Press down on top with other bowl. Remove top bowl and turn out patties onto skillet. It is perfectly formed and did not stick!

**Bean Burgers**: Any cooked and slightly mashed bean, including soybeans, may substitute for the lentils. Flavor and color will vary.

*When properly browned, these "burgers" really look like burgers, texture, color and all. Add some catsup, pickles and a bun, and maybe they'll even pass the taste test!*

---

Per serving (excluding unknown items): 291.3 Calories; 16.2g Fat (49.1% calories from fat); 12.3g Protein; 25.4g Carbohydrate; 72mg Cholesterol; 638mg Sodium. Exchanges: 1 ½ Grain(Starch); 1 Lean Meat; 3 Fat.

# Macaroni Skillet

Makes 6 servings. Prep Time: 1 hour

**2 c elbow macaroni**
**1 med onion, chopped**
**3 tbsp shortening**
**¼ c cornstarch**
**4 c milk, reconstituted**
**2 tsp dill weed**
**2 tsp parsley flakes**
**¼ tsp garlic powder**
**½ tsp pepper**
**½ tsp salt**
**⅓ c bread crumbs**
**paprika, for garnish**
**1 c water**

1. Saute onion in shortening until tender. Remove from heat. Stir in cornstarch mixed with ½ cup water.
2. Blend in milk with a whisk. Cook and stir over medium heat until thick.
3. Add uncooked macaroni and seasonings to the sauce; then stir in water.
4. Cover skillet and cook over low heat 30-40 minutes until macaroni is tender and bubbly.
5. Uncover and top with bread crumbs and paprika. Remove from heat.

Per serving (excluding unknown items): 283.3 Calories; 12.6g Fat (40.0% calories from fat); 8.9g Protein; 33.6g Carbohydrate; 22mg Cholesterol; 314mg Sodium. Exchanges: 1 ½ Grain(Starch); ½ Non-Fat Milk; ½ Vegetable; 2 ½ Fat.

# Marco Polo Spaghetti

Makes 6 servings. Prep Time: 20 minutes

**16 oz spaghetti**
**3 tbsp olive oil**
**salt and pepper, to taste**
**½ tsp garlic powder**
**1 lg green bell pepper, chopped**
**1 lg red bell pepper, chopped**
**¼ c chopped parsley**
**½ c chopped green onions**
**1 can tuna in water, drained**
**½ c chopped nuts, optional**

1. Cook spaghetti according to package directions. Drain. Immediately toss with olive oil, salt, pepper and garlic.
2. Meanwhile, chop and prepare all remaining ingredients.
3. Toss with prepared spaghetti and serve immediately while still hot.

*This recipe came from Julia Child, when she visited Mr. Rogers on his children's program. She said she thought this might have been the way Marco Polo would have experienced spaghetti on his trip to China. They did not have the tomato sauce that we are used to.*

Per serving (excluding unknown items): 450.7 Calories; 14.9g Fat (29.6% calories from fat); 18.0g Protein; 61.5g Carbohydrate; 7mg Cholesterol; 87mg Sodium. Exchanges: 4 Grain(Starch); 1 Lean Meat; ½ Vegetable; 2 ½ Fat.

# Pasta Primavera

Makes 6 servings. Prep Time: 25 minutes

**12 oz spaghetti**
**2 tbsp oil**
**1 med onion**
**½ tsp garlic powder**
**2 c sliced zucchini**
**2 c broccoli flowerets**
**1 c carrots, julienned**
**2 tbsp parsley**
**2 tsp basil**
**½ tsp salt**
**1/8 tsp pepper**
**1/3 c chicken broth**

1. Cook spaghetti until tender and drain in colander.
2. Meanwhile, in a large skillet saute onion over medium-high heat, stirring frequently. Cook 1-2 minutes.
3. Add vegetables; continue stirring and cook 2-3 minutes more.
4. Add broth and seasonings. Simmer 4-5 minutes, or until vegetables are tender-crisp.
5. Toss together with hot spaghetti. Serve immediately.

**Pasta Primavera With White Sauce**: Cook as above. Prepare 3 c White Sauce (see recipe) and stir into cooked vegetables. Milk in the white sauce provides complementary protein with the wheat in the spaghetti.

*This dish should be served with milk or with a side dish using legumes to provide complementary protein with the wheat in the spaghetti.*

---

Per serving (excluding unknown items): 288.0 Calories; 5.8g Fat (18.1% calories from fat); 9.8g Protein; 49.6g Carbohydrate; 0mg Cholesterol; 286mg Sodium. Exchanges: 3 Grain(Starch); 1 Vegetable; 1 Fat.

# Polenta

Makes 8 servings. Prep Time: 1 hour

**2 ½ c chicken broth**
**1 ½ tsp parsley flakes**
**1/8 tsp marjoram**
**dash pepper**
**1 c yellow cornmeal**
**2 tbsp shortening**
**1 c tomato sauce, seasoned**
**2 tbsp shortening, for frying**

1. In a 2-quart pot, combine broth, parsley, marjoram and pepper. Bring to a boil. Add shortening and stir to melt.
2. Whisk in cornmeal to prevent lumping. Reduce heat and continue to cook and stir until mixture is very thick.
3. Grease a 3 ½ x 7 ½ inch loaf pan. Spread polenta evenly in dish. Cool 30 minutes or more, until warm but set.
4. Invert dish. Cut polenta in 16 slices.
5. Fry as for Fried Mush recipe, then arrange slices in a serving dish, overlapping to form a "fan" effect.
6. Pour (warmed, if desired) tomato sauce across slices.

As a main dish, serve with legumes or milk for protein.

Per serving (excluding unknown items): 153.2 Calories; 7.5g Fat (44.1% calories from fat); 5.3g Protein; 16.2g Carbohydrate; 1mg Cholesterol; 675mg Sodium. Exchanges: 1 Grain(Starch); ½ Lean Meat; ½ Vegetable; 1 ½ Fat.

# Red Beans and Rice

Makes 6 servings. Prep Time: 30 minutes

**3 c cooked red kidney beans**
**1 c chopped onion**
**1 med green bell pepper, chopped**
**1 tbsp Worcestershire sauce**
**½ tsp thyme**
**2 tbsp oil**
**½ tsp salt**
**¼ tsp garlic powder**
**⅛ tsp pepper**
**1 c tomato sauce**
**1 tbsp vinegar**
**6 c cooked rice**

1. Saute onion and chopped pepper in oil in a large skillet.
2. Add remaining ingredients except rice. Simmer 10-15 minutes to blend flavors.
3. Serve over hot rice.

**Red Beans And Salsa Over Rice**: Add only beans to sauteed vegetables. Prepare cold salsa using 2 c chopped fresh or canned tomatoes, ¾ c diced onion, ½ tsp garlic powder, 1 T vinegar, 1 tsp oil and 3 dashes Tabasco sauce.

*This is even better with black beans!*

Per serving (excluding unknown items): 426.5 Calories; 5.7g Fat (11.9% calories from fat); 13.8g Protein; 80.0g Carbohydrate; 0mg Cholesterol; 456mg Sodium. Exchanges: 4 ½ Grain(Starch); ½ Lean Meat; 1 Vegetable; 1 Fat.

# Soy Patties

Makes 8 servings. Prep Time: 45 minutes

**2 c soybeans, cooked**
**2 c cooked rice**
**2 tbsp shortening**
**1 med onion, finely chopped**
**½ tbsp soy sauce**
**½ tsp salt**
**¼ tsp garlic powder**
**1 c fine dry breadcrumbs**
**6 tbsp shortening, for frying**

1. Make 2 c soybean puree, as thick as possible, following instructions in Bean section.
2. Thoroughly mix together all ingredients except bread crumbs.
3. Shape into patties (see Lentil Burger for instructions.) Cover with bread crumbs.
4. Bake in a greased pan in a 350 degree oven until brown, or fry in shortening in a skillet on the stove, about 5 minutes per side.

NOTE: Worcestershire sauce may be substituted for soy sauce. Other seasonings may be added, such as beef bouillon granules, sage, etc. Other pureed beans may be substituted for the soybeans.

Serve with gravy or as burgers in buns.

Per serving (excluding unknown items): 310.2 Calories; 17.6g Fat (50.0% calories from fat); 10.4g Protein; 29.0g Carbohydrate; 0mg Cholesterol; 316mg Sodium. Exchanges: 2 Grain(Starch); 1 Lean Meat; ½ Vegetable; 3 Fat.

# Twenty-Minute Tamale Pie

Makes 6 servings. Prep Time: 25 minutes

**1 lg onion, chopped**
**2 c cooked red kidney beans**
**1 pt canned tomatoes**
**2 c canned corn**
**1 c evaporated skim milk**
**1 c cornmeal**
**1 tsp salt**
**1 tbsp chili powder**
**½ tsp cumin**
**2 tbsp shortening**

1. Heat shortening in a large skillet. Add onion and cook until tender.
2. Make evaporated milk from powdered milk by making double-concentrated milk (1 c water with 6 tbsp powdered milk).
3. Add all ingredients to onions and stir until thoroughly mixed. Cover and simmer for 20 minutes, until thickened.

Per serving (excluding unknown items): 314.1 Calories; 5.9g Fat (16.3% calories from fat); 13.1g Protein; 55.5g Carbohydrate; 2mg Cholesterol; 1127mg Sodium. Exchanges: 3 Grain(Starch); ½ Lean Meat; ½ Non-Fat Milk; 1 Vegetable; 1 Fat.

# Spanish Bean Pot

Makes 4 servings. Prep Time: 45 minutes

**2 c cooked garbanzo beans**
**1 tbsp oil**
**½ c chopped onion**
**1 tsp salt**
**½ tsp garlic powder**
**¼ c chopped green bell pepper**
**1 pt canned tomatoes, diced**
**2 c canned corn, reserve liquid**
**½ c green chiles, diced**
**¼ tsp oregano**
**¼ tsp cumin**

1. Saute onion and green pepper in oil in a 3-quart saucepan.
2. Add remaining ingredients and stir well. Cover and simmer 20-30 mins.

**Baked Spanish Bean Pot**: Saute vegetables in skillet. Stir together all ingredients in a 2 ½-quart baking dish. Add liquid from canned corn to barely cover beans. Bake at 300 degrees for about one hour.

Per serving (excluding unknown items): 289.2 Calories; 6.6g Fat (19.1% calories from fat); 11.6g Protein; 51.6g Carbohydrate; 0mg Cholesterol; 1069mg Sodium. Exchanges: 3 Grain(Starch); ½ Lean Meat; 1 ½ Vegetable; ½ Fat.

# Spanish Rice

Makes 6 servings. Prep Time: 45 minutes

**4 tbsp shortening**
**2 c cooked lentils**
**1 med onion, chopped**
**2 c water**
**1 c rice**
**½ c chopped green bell pepper**
**1 pt canned tomatoes**
**1 tsp chili powder**
**½ tsp chili powder**
**1 tsp salt**
**1/8 tsp pepper**

1. In a 10-inch skillet, saute lentils and onion in shortening until onions are tender.
2. Stir in rice to brown slightly.
3. Add water and remaining ingredients. Heat to boiling. Reduce heat, cover and simmer, stirring occasionally until rice is tender, about 30 minutes.

NOTE: Hydrated TVP granules or other cooked beans may be substituted for the cooked lentils.

*This dish may be baked: Use an ungreased 2 ½-quart casserole. Bake at 375 degrees for 45 minutes.*

---

Per serving (excluding unknown items): 294.3 Calories; 9.4g Fat (28.1% calories from fat); 9.3g Protein; 44.4g Carbohydrate; 0mg Cholesterol; 538mg Sodium. Exchanges: 2 ½ Grain(Starch); ½ Lean Meat; 1 Vegetable; 1 ½ Fat.

# Spanish Wheat

Makes 6 servings. Prep Time: 1 hour and 15 minutes

**4 tbsp oil**
**2 c cooked lentils**
**1 med onion, chopped**
**2 c water**
**1 c whole wheat berries**
**½ c chopped green bell pepper**
**1 pt canned tomatoes**
**1 tsp chili powder**
**½ tsp oregano**
**1 tsp salt**
**1/8 tsp pepper**

1. In a 10-inch skillet, saute lentils and onion in shortening until onions are tender. Remove from pan.
2. Add water and wheat. Cover, reduce heat and simmer about 45 minutes, until wheat is tender and kernels begin to pop open, adding water during cooking if needed.
3. Return onion mixture to pan and add remaining ingredients and simmer another 30 minutes so wheat absorbs the flavor. Simmer uncovered, if needed for extra moisture to evaporate.

*2 c hydrated TVP granules may be used instead of lentils.*

Per serving (excluding unknown items): 266.1 Calories; 10.0g Fat (32.1% calories from fat); 10.0g Protein; 37.4g Carbohydrate; 0mg Cholesterol; 539mg Sodium. Exchanges: 2 Grain(Starch); ½ Lean Meat; 1 Vegetable; 2 Fat.

# Speedy Spaghetti

Makes 6 servings. Prep Time: 40 minutes

**1 c textured soy protein**
**1 c beef broth, from base**
**1 qt canned tomatoes**
**¾ c chopped green bell pepper**
**1 lg onion, chopped**
**3 tbsp oil**
**2 tsp salt**
**1 tsp sugar**
**1 tsp basil**
**½ tsp oregano**
**½ tsp marjoram**
**10 ozs spaghetti, broken up**

1. Soak TVP in beef broth until broth is absorbed. Cook and stir TVP, onion and green pepper in oil in a 10-inch skillet or 4-quart pot until onion is tender.
2. Add remaining ingredients, breaking up tomatoes with a fork.
3. Heat to boiling. Then reduce heat, cover and simmer.
4. Stir occasionally and cook until spaghetti is tender, about 30 minutes.

*This is Speedy because the spaghetti does not need to be precooked! Beans or lentils may be substituted for TVP and tomato sauce may be substituted for canned tomatoes.*

Per serving (excluding unknown items): 394 Calories; 9.8g Fat (11.4% calories from fat); 47.5g Protein; 51.2g Carbohydrate; 0mg Cholesterol; 683mg Sodium.
Exchanges: 2 ½ Grain(Starch); 3 ½ Lean Meat; 2 Vegetable; 1 ½ Fat.

# Stir Fry Dinner

Makes 6 servings. Prep Time: 40 minutes

**6 c cooked rice**
**2 c soft cheese, cut in ½" cubes**
**¼ c soy sauce**
**1 ½ c chicken broth, from base**
**2 tbsp cornstarch**
**¼ c cold water**
**6 c mixed vegetables**

1. Prepare rice as directed on package. Meanwhile, marinate cheese in soy sauce.
2. Cut up vegetables, any kind available, in slices or slivers so that the firmer ones are in smaller pieces and all will get tender in about the same cooking time.
3. In a large skillet or wok, heat oil to medium high. Drain cheese (reserve soy sauce) and brown in oil several minutes; then remove to plate.
4. Stir fry vegetables until crisp-tender. Return cheese to skillet.
5. Add remaining soy sauce and broth and heat to boiling. Meanwhile, stir together cornstarch and cold water. When broth is boiling, stir in liquified cornstarch. Cook, stirring constantly, until liquid is thick and clear, 2-3 minutes.
6. Pour mixture over hot rice and serve immediately.

**Stir Fry Theme And Variations**: Spaghetti or Chinese noodles may be substituted for the rice. Any type of cooked bean or lentil may be substituted for the tofu. Beef broth may be substituted for the chicken broth. Any combination of vegetables may be used. The possibilities are endless!

---

Per serving (excluding unknown items): 410.1 Calories; 2.0g Fat (4.5% calories from fat); 20.3g Protein; 75.3g Carbohydrate; 4mg Cholesterol; 1636mg Sodium. Exchanges: 3 ½ Grain(Starch); 1 ½ Lean Meat; 3 ½ Vegetable.

# Tuna Patties

Makes 6 servings. Prep Time: 45 minutes

**1 c whole wheat flour**
**½ tsp salt**
**2 tsp baking powder**
**1 lg egg, reconstituted**
**½ c milk, reconstituted**
**2 c tuna in water**
**6 tbsp shortening, for frying**

1. Combine all ingredients and shape into 6 patties. (See Lentil Burger for instructions of forming patties.)
2. Roll in fine bread crumbs if making traditional croquettes. Leave plain if making tuna patties.
3. Brown in hot shortening or oil, about 5 minutes each side.

*Mashed beans or lentils, finely chopped vegetables or soaked TVP granules may be substituted for the tuna.*

Per serving (excluding unknown items): 265.6 Calories; 15.1g Fat (50.5% calories from fat); 17.4g Protein; 15.9g Carbohydrate; 53mg Cholesterol; 492mg Sodium. Exchanges: 1 Grain(Starch); 2 Lean Meat; 3 Fat.

# Rice Patties

Makes 6 servings. Prep Time: 45 minutes

**1 ½ c cooked rice**
**2 lg eggs, reconstituted**
**¾ c dry bread crumbs**
**½ c sliced green onions**
**¼ tsp salt**
**½ tsp sage**
**4 tbsp shortening, for frying**

1. Combine all ingredients thoroughly. Shape into 6 half-inch thick patties. (See Lentils Burgers for instructions of forming patties.)
2. Heat oil or shortening in a large skillet over medium heat. Fry until golden brown, 3-4 minutes per side.

Per serving (excluding unknown items): 219.4 Calories; 11.3g Fat (46.5% calories from fat); 5.1g Protein; 24.0g Carbohydrate; 72mg Cholesterol; 206mg Sodium. Exchanges: 1 ½ Grain(Starch); ½ Lean Meat; 2 Fat.

# Wheat Verde

Makes 4 servings. Prep Time: 45 minutes

**2 tbsp shortening**
**1 med onion, chopped**
**1 c cracked wheat**
**3 med green chiles, chopped**
**1 tbsp chicken bouillon granules**
**1/8 tsp pepper**
**¼ tsp cumin**
**½ tsp garlic powder**
**1 tsp oregano**
**2 c water**
**¼ tsp salt**

1. In large skillet, saute onion in shortening until transparent.
2. Add cracked wheat and fry over low heat, 3-5 minutes, to soak fat into each kernel of wheat.
3. Stir in remaining ingredients. Cook about 25 minutes over low heat, stirring frequently.
4. Wheat should be soft, but grains separated and barely moist.

**Beef Flavored**: Substitute beef bouillon granules for the chicken.

**Stir-Fried Cracked Wheat**: Omit chiles, cumin and oregano. 2 tbsp soy sauce may be added.

*"Verde" is Spanish for "green," referring to the green chiles in this dish.*

---

Per serving (excluding unknown items): 213.9 Calories; 7.5g Fat (29.8% calories from fat); 6.0g Protein; 34.0g Carbohydrate; 0mg Cholesterol; 634mg Sodium. Exchanges: 2 Grain(Starch); 1 Vegetable; 1 ½ Fat.

# BAKED MAIN DISHES

If you have an oven available, you have come to the right place. These recipes include many foods from Mexico and Italy, among other places. Most of them will be reminiscent of familiar recipes, but they will have off-the-grid adaptations: replacements for meat, cheese, cottage cheese, etc. Some recipes require pre-cooked beans or rice.

To save time and fuel, some recipes have been adapted to used uncooked pasta or rice: Lazy Lasagna, Herbed Lentils and Rice, Macaroni Bake, Six-Layer Casserole and Texas Hash.

If you are baking a batch of bread, consider doubling it and using the extra in Calzones or Vegetarian Pizza. Or perhaps you may use it to make some French Bread to go along with the Lasagna, or make some rolls for one of the other casseroles.

Several bean pot, or baked bean, recipes are included. Since they use precooked beans, these recipes are a good way of recycling leftover beans from a previous meal. Bean pots typically require long, slow baking. If you are using your baking appliance to also heat your house, maybe today is a good day to try a bean pot.

# Baked Lentils

Makes 4 servings. Prep Time: 1 hour and 15 minutes

**3 c cooked lentils**
**1 tbsp oil**
**½ c chopped onion**
**1 tsp salt**
**1 c canned tomatoes**
**⅓ c chopped green bell pepper**
**2 med green onions, chopped**

1. Dice tomatoes. Combine all ingredients and enough liquid (water or bean cooking liquid) to barely cover lentils in a 1 ½-quart baking dish.
2. Bake, uncovered, at 300 degrees for 1 hour.

Per serving (excluding unknown items): 248.4 Calories; 4.3g Fat (14.7% calories from fat); 15.6g Protein; 40.3g Carbohydrate; 0mg Cholesterol; 667mg Sodium. Exchanges: 2 Grain(Starch); 1 Lean Meat; 2 Vegetable; ½ Fat.

# Baked Potato Toppers

Makes 6 servings. Prep Time: 1 hour

**6 lg baking potatoes**
**3 c lentil chili**
**3 c chili**
**3 c white sauce, with cooked vegetables**
**3 c spaghetti sauce, with TVP or lentils**
**3 c Ratatouille**

1. Scrub potatoes thoroughly and prick with a fork.
2. Bake potatoes in a 350 degree oven for 45-60 minutes, until quite tender.
3. Choose one of the above toppings, using leftovers from a previous meal.
4. Reheat topping in a saucepan.
5. To serve, split open potato and pour about ½ c topping into each potato.

Per serving (excluding unknown items): 261.3 Calories; 7.2g Fat (24.8% calories from fat); 6.5g Protein; 42.2g Carbohydrate; 22mg Cholesterol; 675mg Sodium. Exchanges: 2 ½ Grain(Starch); ½ Lean Meat; 1 Fat.

# Bean and Potato Scallop

Makes 4 servings. Prep Time: 1 hour and 30 minutes

**3 c cooked red kidney beans**
**1 lg sliced onion**
**½ tsp salt pepper, to taste**
**4 tbsp shortening**
**1 tbsp cornstarch**
**½ tsp salt**
**2 ½ c milk, reconstituted**
**3 lg potatoes, sliced**

1. Preheat oven to 350 degrees. In a medium skillet, saute onion in 2 tbsp shortening until tender. Add beans, salt and pepper; then remove from skillet.
2. Melt 2 tbsp more shortening in skillet. Stir in cornstarch and salt until smooth. Gradually add milk, stirring with a whisk until smooth. Stir constantly over medium heat until mixture thickens and comes to a boil.
3. In a 3-quart casserole, layer ⅓ of the potato slices and half the bean mixture. Pour about a third of the sauce on top.
4. Repeat layers, ending with potato slices and topping with remaining sauce.
5. Bake, uncovered, until potato slices are tender and crusty brown on top, about one hour.

**Plain Potato Scallop**: Omit beans. Add 3 additional sliced potatoes.

Per serving (excluding unknown items): 609.3 Calories; 18.9g Fat (27.3% calories from fat); 22.5g Protein; 90.7g Carbohydrate; 21mg Cholesterol; 628mg Sodium. Exchanges: 5 Grain(Starch); ½ Lean Meat; ½ Non-Fat Milk; ½ Vegetable; 3 ½ Fat.

# Boston Baked Beans

Makes 6 servings. Prep Time: 2 hours

**6 c cooked white beans**
**1 med onion, sliced**
**½ c molasses**
**3 tbsp sugar**
**1 tsp dry mustard**
**1 ½ tsp salt**
**¼ tsp pepper**
**2 tbsp oil**

1. Saute onion in oil until tender. Combine all ingredients in a 2 ½-quart bean pot (baking dish) with a cover.
2. Add enough water or cooking liquid to barely cover beans. Cover and bake about 2 hours, until thickened.

**Southern Style Bean Pot**: Use beans and onions as above. Substitute the following seasonings. 1 ½ tsp salt, 1 bay leaf, 1 cayenne pepper (opt), ¼ c molasses, $^{1}/_{3}$ c catsup, 1 tbsp prepared mustard, ½ tsp ginger, 1 tbsp Worcestershire sauce, and 2 tbsp brown sugar.

*Baked beans could use lots of cooking fuel, with the precooking of the beans and then a long baking time. But if you are using leftover beans from a large batch, and if your baking appliance is also heating your home (such as a wood cookstove or a wood heating stove with a stovetop oven), then a bean pot costs you nothing extra!*

---

Per serving (excluding unknown items): 396.6 Calories; 5.3g Fat (11.7% calories from fat); 17.8g Protein; 72.2g Carbohydrate; 0mg Cholesterol; 555mg Sodium. Exchanges: 3 Grain(Starch); 1 ½ Lean Meat; ½ Vegetable; ½ Fruit; 1 Fat; 1 ½ Other Carboh

## Boston Baked Wheat

Makes 4 servings. Prep Time: 45 minutes

**4 c cooked whole wheat berries**
**1 lg chopped onion**
**1 tbsp oil**
**½ c molasses**
**1 tsp prepared mustard**
**1 c catsup**
**salt and pepper, to taste**
**½ c boiling water**

1. Cook onion in oil in skillet over medium heat until softened.
2. Combine onion and remaining ingredients in a 2 ½-quart deep baking dish.
3. Bake at 325 degrees for 30-45 minutes, until thickened.

Per serving (excluding unknown items): 367.6 Calories; 4.2g Fat (9.7% calories from fat); 7.0g Protein; 81.7g Carbohydrate; 0mg Cholesterol; 754mg Sodium. Exchanges: 2 Grain(Starch); ½ Vegetable; ½ Fat; 3 Other Carbohydrates.

## Herbed Lentils and Rice

Makes 4 servings. Prep Time: 1 hour and 15 minutes

**3 c chicken broth, from base**
**¾ c lentils, uncooked**
**½ c rice, uncooked**
**¾ c chopped onion**
**¼ tsp salt**
**½ tsp basil**
**¼ tsp oregano**
**¼ tsp garlic powder**
**⅛ tsp pepper**

1. Combine all ingredients in an ungreased 1 ½-quart casserole.
2. Bake, covered, in a 350 degree oven for 1-1 ½ hours, or until lentils and rice are tender and moisture is absorbed. Stir twice during cooking, if possible.

Per serving (excluding unknown items): 277.1 Calories; 2.5g Fat (8.1% calories from fat); 20.5g Protein; 43.4g Carbohydrate; 2mg Cholesterol; 1313mg Sodium. Exchanges: 2 ½ Grain(Starch); 2 Lean Meat; ½ Vegetable.

# Calzones

Makes 8 servings. Prep Time: 45 minutes

**2 lb bread dough**
**2 c soft cheese**
**1 tsp salt**
**1 tsp parsley flakes**
**½ tsp basil**
**1 tsp oregano**
**½ tsp garlic powder**
**1 pt tomato sauce**
**½ tsp oregano**
**1 ½ tsp parsley**
**¼ tsp garlic**

1. Divide dough into 8 equal portions and roll out into rounds about ¼-inch thick. Preheat oven to 450 degrees.
2. Combine cheese, salt, parsley, basil, oregano, and garlic with a fork.
3. Fill each dough round with about ½ c cheese filling, placing it on half the circle and leaving a ½-inch rim.
4. Moisten the rim with water, fold empty side over and crimp edge with a fork, being careful not to break the crust.
5. Bake on a lightly greased baking sheet for 15-20 minutes or until crisp and lightly browned.
6. Serve warm, topped with warmed tomato sauce, or may be served cold like a sandwich.

**Bean-Filled Calzones**: Omit cheese filling. Use 2 ½ c slightly mashed pinto beans; 1 onion, chopped and sauteed in 1 tbsp oil/ ½ tsp garlic powder; ½ tsp basil; ½ tsp salt; and 1/8 tsp crushed fennel seed.

**Tuna Calzones**: Instead of cheese filling, use 2 cans tuna, 1 reconstituted egg, ½ c fine bread crumbs, ½ c chopped onion, 2 tbsp chopped parsley, ½ tsp salt and 1/8 tsp pepper.

*"Calzone" is the word "trousers" in Italian. I'm not sure where the name came from, but these taste lots better than pants!*

Per serving (excluding unknown items): 412.1 Calories; 15.6g Fat (33.6% calories from fat); 14.2g Protein; 55.1g Carbohydrate; 2mg Cholesterol; 1897mg Sodium. Exchanges: 3 ½ Grain(Starch); 1 Lean Meat; 1 Vegetable; 3 Fat.

# Cheese Enchiladas

Makes 6 servings. Prep Time: 1 hour

**2 c soft cheese**
**2 c cooked spinach**
**8 med green onions, finely chopped**
**2 c enchilada sauce**
**2 c tomato sauce**
**12 lg tortillas**

1. Preheat oven to 350 degrees. Shred cheese and chop spinach.
2. Blend together cheese, spinach and green onions.
3. Mix sauces together and heat in saucepan.
4. To make enchiladas, dip tortillas into warm sauce to soften. Spoon about 1/3 cup prepared filling into center. Roll up tortillas by folding in sides 1 inch and then rolling front to back.
5. Place in 9x13 baking dish. Pour remaining sauce over the top. Bake uncovered 30 minutes.

**Bean Enchiladas**: Substitute 2 cups refried beans or cooked beans for either the cheese or the spinach.

Per serving (excluding unknown items): 464.3 Calories; 14.3g Fat (26.6% calories from fat); 22.0g Protein; 66.8g Carbohydrate; 28mg Cholesterol; 971mg Sodium. Exchanges: 2 ½ Grain(Starch); 2 Lean Meat; 5 Vegetable; 2 ½ Fat.

# Lazy Lasagna

Makes 6 servings. Prep Time: 1 hour and 20 minutes

**12 oz lasagna noodles**
**1 lb spinach leaves**
**1 qt tomato sauce, seasoned**
**1 c water**
**2 c soft cheese**
**1 tsp salt**
**1 tsp parsley flakes**
**½ tsp basil**
**½ tsp oregano**
**½ tsp garlic powder**

1. Preheat oven to 350 degrees. Stir together sauce and water. Lightly steam and chop spinach.
2. Stir together cheese, spinach and seasonings.
3. Assemble ingredients in a greased 9x13 casserole in the following order:
   - small amount of tomato sauce
   - 1 layer uncooked noodles
   - 1/3 of the cheese mixture
   - ¼ of the tomato sauce
4. Repeat as above, ending with noodles and sauce.
5. Cover tightly and bake 45 minutes. Uncover and bake 15 minutes more, until excess moisture is absorbed and noodles are tender. Let stand 15 minutes before serving.

**Bean Lasagna**: Omit spinach. Add 2 cups cooked red beans, pinto beans or lentils to tomato sauce before assembling lasagna.

**"Poor Boy" Lasagna**: Substitute 12 oz uncooked large pasta (rotelle, ziti, rigatoni, etc) for lasagna noodles. Mix together pasta, sauce and filling rather than layering.

*Lazy because you don't have to precook the noodles!*

---

Per serving (excluding unknown items): 318.4 Calories; 1.7g Fat (4.5% calories from fat); 20.0g Protein; 57.9g Carbohydrate; 3mg Cholesterol; 1415mg Sodium. Exchanges: 3 Grain(Starch); 1 Lean Meat; 2 ½ Vegetable.

# Macaroni Bake

Makes 6 servings. Prep Time: 45 minutes

**1 ½ c elbow macaroni, uncooked**
**1 can peas, undrained**
**1 can tuna in oil, drained**
**2 ½ c white sauce, thin**

1. Combine all ingredients in a 2-quart casserole.
2. Bake, covered, in a 350 degree oven for 35-45 minutes, until macaroni are tender and liquid is absorbed.

*This is so easy because the macaroni requires no pre-cooking! Any vegetable may be substituted for the peas. One can cream of mushroom soup and one cup reconstituted milk may be substituted for the white sauce.*

Per serving (excluding unknown items): 393.5 Calories; 18.1g Fat (40.3% calories from fat); 16.0g Protein; 44.5g Carbohydrate; 5mg Cholesterol; 2126mg Sodium. Exchanges: 3 Grain(Starch); 1 Lean Meat; 3 Fat.

# Texas Hash

Makes 6 servings. Prep Time: 1 hour

**3 tbsp oil**
**3 lg onions, chopped**
**1 lg chopped green bell pepper**
**½ c textured soy protein**
**½ c chicken broth**
**1 qt canned tomatoes**
**2 c elbow macaroni, uncooked**
**1 tsp chili powder**
**½ tsp pepper**
**½ tsp salt**
**½ c water**

1. Preheat oven to 350 degrees.
2. In a large (2 or 3-quart) oven-safe skillet, heat oil. Saute onions and green pepper to soften.
3. Add remaining ingredients and stir.
4. Cover and bake 30 minutes. Remove cover and bake an additional 15-20 minutes, until liquid is absorbed and macaroni is tender.

Per serving (excluding unknown items): 261.1 Calories; 8.1g Fat (26.1% calories from fat); 18.7g Protein; 32.8g Carbohydrate; 0mg Cholesterol; 660mg Sodium. Exchanges: 1 Grain(Starch); 2 Lean Meat; 2 ½ Vegetable; 1 ½ Fat.

# Red Beans and Cornbread

Makes 6 servings. Prep Time: 1 hour

---

**1 c yellow cornmeal**
**1 ½ c whole wheat flour**
**1 tbsp sugar**
**1 tbsp baking powder**
**3 tbsp soy flour**
**1 ¾ c water**
**¼ c powdered milk**
**2 tbsp oil**
**1 c canned corn**
**1 tsp salt**
**½ c red and green bell peppers, chopped**
**½ c chopped onion**
**1 tbsp oil**
**2 c kidney beans, cooked**
**1 c tomato sauce**
**1 tsp chili powder**
**1 tsp Worcestershire sauce**

1. Make cornbread by stirring together add dry ingredients. Combine water and oil.
2. Add all at once to dry ingredients and beat just until smooth. Fold in corn and chopped peppers.
3. Turn into a greased 9x9 baking pan. Bake in a 400 degree oven, 35-40 minutes, until golden.
4. Meanwhile, cook onion in oil (or shortening) until tender.
5. Add remaining ingredients and simmer, uncovered, about 8 minutes or until heated through. Mash beans slightly.
6. Cut cornbread into 6 rectangles. Spoon bean mixture on top of each square.

---

Per serving (excluding unknown items): 417.0 Calories; 10.4g Fat (21.5% calories from fat); 15.1g Protein; 70.3g Carbohydrate; 5mg Cholesterol; 913mg Sodium. Exchanges: 4 Grain(Starch); ½ Lean Meat; ½ Vegetable; 2 Fat.

# Six-Layer Casserole

Makes 6 servings. Prep Time: 1 hour and 15 minutes

**1 c rice**
**2 c canned corn**
**1 tsp salt**
**¼ tsp pepper**
**1 tsp beef bouillon granules**
**1 pt tomato sauce**
**¾ c boiling water**
**1 tsp Worcestershire sauce**
**1 tsp oregano**
**½ tsp basil**
**1 c chopped onion**
**½ c chopped green bell pepper**
**1 c textured soy protein, granular**
**1 c beef broth, from base**
**½ c bread crumbs**

In a 2 ½-quart greased casserole, layer as follows:

1. Rice mixed well with corn, half the salt and pepper, bouillon and water.
2. Half the tomato sauce with half the herbs and Worcestershire.
3. Chopped onion and green pepper.
4. TVP, remaining salt and pepper.
5. Remaining tomato sauce, herbs and Worcestershire.
6. Sprinkle bread crumbs on top. Cover tightly and bake at 375 degrees for 45 minutes. Uncover and bake an additional 15 minutes.

*Two c cooked beans or lentils may be substituted for the TVP.*

---

Per serving (excluding unknown items): 371.6 Calories; 1.8g Fat (4.1% calories from fat); 34.0g Protein; 62.4g Carbohydrate; 0mg Cholesterol; 1440mg Sodium. Exchanges: 3 ½ Grain(Starch); 3 ½ Lean Meat; 1 ½ Vegetable.

# Tamale Pie

Makes 6 servings. Prep Time: 1 hour and 20 minutes

**3 tbsp oil**
**½ c textured soy protein, granular**
**½ c beef broth, from base**
**¾ c chopped onion**
**½ tsp garlic powder**
**½ c chopped green bell pepper**
**2 c canned corn**
**2 c red kidney beans, cooked**
**2 c tomato sauce**
**2 tsp chili powder**
**3 tbsp Salsa**
**1 c yellow cornmeal**
**2 ½ c water**
**¼ tsp salt**

Preheat oven to 375 degrees. Soak TVP in beef broth until hydrated.
2. Heat oil in large skillet. Add onion and TVP, cooking until onion is tender. Add garlic, corn, beans, tomato sauce, chili powder and salsa. Mix thoroughly.
3. Pour into a greased 10x10 or 8x11 baking pan.
4. Make topping by combining cornmeal, water and salt in the same skillet, using a whisk to prevent lumping.
5. Stir constantly, bringing mixture to a boil, and continue cooking until slightly thickened. Spoon on top of bean mixture.
6. Bake 45-60 minutes, until topping is dry and firm.

**Stovetop Tamale Pie**: This casserole may be "baked" on top of the stove by cooking slowly in a skillet with a tight lid, for about 30-45 minutes. Uncover last few minutes to allow cooked topping to dry out.

---

Per serving (excluding unknown items): 377.2 Calories; 8.4g Fat (18.8% calories from fat); 24.1g Protein; 58.0g Carbohydrate; 0mg Cholesterol; 910mg Sodium. Exchanges: 3 Grain(Starch); 2 Lean Meat; 1 ½ Vegetable; 1 ½ Fat.

# Vegetarian Pizza

Makes 6 servings. Prep Time: 1 hour

**2 tbsp yeast**
**1 ½ c warm water, 105-115 degrees**
**1 ½ tsp sugar**
**1 ½ tsp salt**
**3 tbsp oil**
**4 c whole wheat flour**
**2 c tomato sauce**
**finely chopped vegetables**
**granular TVP, lentils, beans or cheese**

1. Sprinkle yeast and sugar over warm water. Let stand 5 minutes, until bubbly.
2. Stir in salt, oil and half the flour; beat vigorously one minute.
3. Add most of remaining flour and stir, adding just enough flour to make dough stiff enough to leave sides of the bowl. Let rest 5 minutes.
4. Divide dough in half, for two pizzas. Pat each half into a circle on oiled 12-inch pizza pan.
5. Spread 1 c sauce on each pizza. Sprinkle with vegetable and protein toppings, as desired. Cheese can be shredded or finely chopped before sprinkling on pizza.
6. Bake in a 425 degree oven until dough is browned, 20-25 minutes.

**Personal Pizzas**: Divide dough into 12 portions. Form each into a 5-6-inch circle. Place on greased cookie sheet. Cover with sauce and toppings. Bake 15-20 minutes at 425 degrees.

**Quick Pizzas**: Use already-made split English muffins or pita bread rounds instead of above pizza dough. Bake just till bread begins to brown, about 10 minutes.

Per serving (excluding unknown items): 371.8 Calories; 8.6g Fat (19.5% calories from fat); 13.6g Protein; 66.5g Carbohydrate; 0mg Cholesterol; 1035mg Sodium. Exchanges: 4 Grain(Starch); 1 Vegetable; 1 ½ Fat.

# Vermont Bean Pot

Makes 6 servings. Prep Time: 2 hours and 30 minutes

**5 c cooked white beans**
**1 ½ c water**
**¼ c soy bacon bits**
**½ c chopped onion**
**2 tbsp oil**
**1 ½ tsp salt**
**1/8 tsp pepper**
**1 tsp ginger**
**½ tsp allspice**
**2 tbsp vinegar**
**2/3 c maple syrup, see recipe**

1. Saute onion in oil in skillet until transparent. Add onion to beans.
2. Combine water, spices and maple syrup in the skillet. Bring to a boil.
3. In a 2-quart casserole with a lid, combine beans and boiled syrup. Stir in bacon bits (or ham-flavored TVP).
4. Cover and bake at 300 degrees for 2 hours. Uncover, bake 30 mins more.

---

Per serving (excluding unknown items): 390.3 Calories; 7.7g Fat (17.3% calories from fat); 17.9g Protein; 65.6g Carbohydrate; 0mg Cholesterol; 722mg Sodium. Exchanges: 2 ½ Grain(Starch); 1 ½ Lean Meat; 1 Fat; 1 ½ Other Carbohydrates.

# Western Beans

Makes 6 servings. Prep Time: 1 hour and 15 minutes

**¼ c soy bacon bits**
**1 med onion, sliced**
**2 tbsp oil**
**½ tsp garlic powder**
**¼ c molasses**
**2 tbsp brown sugar**
**2 tbsp vinegar**
**2 tsp prepared mustard**
**2 c cooked pinto beans**
**2 c cooked kidney beans**
**2 c baked beans**

1. NOTE: Any combination of cooked beans available may be used, totaling 6 c. Baked beans may be canned.
2. Saute onion in oil in a skillet until tender.
3. Combine all in a 2 ½-quart casserole. Cover and bake one hour.

---

Per serving (excluding unknown items): 387.2 Calories; 8.2g Fat (18.0% calories from fat); 17.9g Protein; 65.9g Carbohydrate; 0mg Cholesterol; 542mg Sodium. Exchanges: 3 Grain(Starch); 1 Lean Meat; ½ Vegetable; 1 Fat; 1 Other Carbohydrates.

# STOVETOP BREADS

Cooking bread on top of the stove is probably older than baking, and in this chapter you will find some very old and basic recipes. Hoe Cakes, Ash Cakes, Corn Pone and Hush Puppies all use cornmeal, a staple food of early America. Stovetop cooking of bread involves one of two processes: frying or steaming.

## Fried Breads

The first section of this chapter lists the frying recipes. Some of these use oil and others are dry-fried. Tortillas, being quite thin, can cook in just a few minutes and can be dry-fried. English muffins are also in this category. However, because they are thicker and take longer to cook, they should be fried at a lower temperature to prevent burning on the outside before they are cooked to the center. Cornmeal in the skillet prevents sticking.

Other breads of medium thickness require a few tablespoons of oil in the pan. These include the Navajo Fry Bread, Irish Soda Scone, Hoe Cakes and Panned Bread. Here again, it will take some experience to get the stove temperature just right: hot enough to leaven the dough and cook it to the center, but not too hot that it burns on the bottom before the center is cooked.

The rest of the recipes (Hard-Shell Tacos, Cake Doughnuts and Hush Puppies) are deep fried, meaning one-inch deep oil or more. A specific temperature for the oil is given, and if there is no thermometer available, you may use the bread cube test for an estimate. With some practice, you will get a feel for the oil temperature as you are cooking. Take care with hot oil! Whereas boiling water reaches 212 degrees, this oil is 375 degrees!

## Steamed Breads

Steamed breads may be cooked in a 6-quart pot or in a pressure cooker. In both cases, a small amount of water is placed in the pot and the bread pan is placed on top of a rack in the pot. The bread pan is covered tightly (with foil) and the pot is also covered with its lid. The pot is kept simmering on the stove and the steam heat cooks the bread. Using a pressure cooker will cut the time by one-half to two-thirds. For example, if bread takes 2-3 hours in a regular pot, it will take only 1 – 1 ½ hours in the pressure cooker. Since the time is much longer without pressure cooking, additional water may need to be added during cooking so the pot does not boil dry. Check several times during cooking. Be sure to follow the instruction booklet provided with your pressure cooker for steaming. The instructions in the recipes are only general in nature and may not apply to your particular pressure cooker.

Sometimes these breads are referred to as "puddings" because they are so moist. Their preparation requires the use of various sized molds. These may be empty food cans, one- or two-quart aluminum or metal molds, or ovenproof glass bowls. Coppertone aluminum molds may also be used. Since it takes time for heat to penetrate to the center of a mold, larger molds with a tube hole in the middle or narrow containers will work best. A large can, such as a #10 can, may not allow the dough to cook in the center. Several 16- or 20-ounce cans would work better.

You may improvise a steamer rack for your steamer pot by balancing a wire rack on top of 3 or 4 inverted custard or coffee cups on the bottom of the pot. If rack is several inches high, water should be added to the level of the rack. If rack is close to the bottom of the pot, water should be added to reach one-third to one-half up the sides of the mold.

## Cleaning Your Steamer

Frequent use of your steamer (either regular or pressure cooker) may result in water deposits or stains inside the pot. These may be prevented by using 1 teaspoon vinegar or ½ teaspoon cream of tartar in the water when steaming.

## Reheating Leftovers

Foods such as casseroles, refried beans or rice may stick to saucepans when reheated. Using your steamer will avoid messy cleanup problems. Just place leftovers in a mold on rack in cooker over one cup water. In a pressure cooker, set the control at 15, cook on high and remove from heat when control begins to jiggle. Let pressure drop normally. If using a regular pot, use two cups water, bring to a boil, reduce heat and let steam for about 20-30 minutes or until food is heated through.

# Croutons

Makes 8 servings. Prep Time: 1 hour and 20 minutes

**¼ tsp garlic powder**
**½ c oil**
**2 c bread cubes, cut in ½" cubes**

1. Combine garlic powder and oil. Let sit one hour.
2. Meanwhile, remove crusts from stale bread and cut into cubes.
3. Place 4 tbsp oil mixture in medium skillet. Saute bread cubes until brown.
4. Keep remaining oil for salad dressing.

**Oven Croutons**: Toss bread cubes with oil. Toast in 350 degree oven for about 10 minutes, stirring occasionally until brown.

*Don't throw that stale bread to the birds! These are great for soups or salads.*

Per serving (excluding unknown items): 150.4 Calories; 14.0g Fat (83.0% calories from fat); 1.0g Protein; 5.5g Carbohydrate; 0mg Cholesterol; 65mg Sodium. Exchanges: 3 Fat.

# Corn Tortillas

Makes 4 servings. Prep Time: 45 minutes

**2 c corn flour**
**½ c water**
**½ tsp salt**
**2 tbsp shortening**

1. Mix ingredients together and knead well. Adjust water as necessary to make workable dough.
2. Let stand for 10 minutes. Divide into 4 equal portions.
3. Knead, pat or slap into thin pancake, 8-10 inches in diameter, on a floured surface.
4. Cook on stovetop in ungreased cast iron or non-stick skillet, turning to cook through but be careful not to burn.
5. See Flour Tortillas recipe for making hard taco shells and tortilla chips.

*1 c corn meal and 1 c all-purpose or soft white wheat flour may be substituted for the corn flour. Corn flour is made by milling whole corn on a fine setting.*

Per serving (excluding unknown items): 211.2 Calories; 2.3g Fat (9.4% calories from fat); 4.1g Protein; 45.0g Carbohydrate; 0mg Cholesterol; 270mg Sodium. Exchanges: 3 Grain(Starch).

# Cake Doughnuts

Makes 10 servings. Prep Time: 45 minutes

**3 1/3 c all-purpose flour**
**¾ c sugar**
**1 tbsp baking powder**
**½ tsp salt**
**2 tbsp shortening**
**½ tsp cinnamon**
**¼ tsp nutmeg**
**3 tbsp soy flour**
**3 tbsp powdered milk**
**1 c water**
**oil, for frying**

1. Heat oil, 2-3 inches deep in deep fat fryer or heavy pot to 375 degrees. NOTE: If you have no thermometer, test the oil as it is heating. A 1-inch cube of bread will brown in about 20-25 seconds in 375 degree oil. After a while, you will get the "feel" of the temperature by how the doughnuts are cooking.
2. Beat 1 ½ c of the flour and all other ingredients until moistened. Continue beating 2 more minutes.
3. Stir in remaining flour. Turn out onto floured surface.
4. Roll gently until 3/8 inch thick. Cut with floured doughnut cutter, or cut by hand with a knife.
5. Slide carefully into hot oil with spatula. Turn doughnuts as they rise to the surface. Fry until golden brown, about 1 minute per side.
6. Remove from oil with tongs or fork. Drain on brown paper or paper towels.
7. Serve plain, sugared or frosted. Makes 18-20 doughnuts and holes.

*Soft white wheat flour or hard red wheat flour will work in this recipe, but as the flour gets heavier, so does the doughnut. But they are so fun to make, it may not matter at all!*

---

Per serving (excluding unknown items): 252.4 Calories; 4.0g Fat (14.1% calories from fat); 5.5g Protein; 48.7g Carbohydrate; 2mg Cholesterol; 226mg Sodium. Exchanges: 2 Grain(Starch); 1 Fruit; ½ Fat; 1 Other Carbohydrates.

# Dumplings

Makes 4 servings. Prep Time: 25 minutes

**2 c whole wheat flour**
**1 tsp salt**
**4 tsp baking powder**
**$^1/_8$ tsp pepper**
**$^2/_3$ c water**
**¼ c powdered milk**
**1 lg egg, reconstituted**
**3 tbsp oil**

1. Stir together dry ingredients, including powdered milk.
2. Add liquids and mix gently, just til blended.
3. Drop by tbsp into boiling liquid ( soup, stew or stewed fruit). Cover and cook 15-20 minutes.

**Herbed Dumplings**: Add $^1/_8$ tsp sage, 2 tbsp parsley flakes.

**Mexican Dumplings**: Add 1 tsp chili powder, ½ tsp cumin, ½ tsp oregano.

**Sweet Dumplings**: Add ¼ c sugar to the above recipe, omit pepper, and cook in stewing plums, other soft fruit or even applesauce.

**Eggless Dumplings**: Increase water to 1 c. Add 1 ½ tbsp soy flour. Omit reconstituted egg.

*Dumplings are an old-fashioned way of baking your biscuits while you are cooking your stew. They are moist and tender, sure to be enjoyed.*

---

Per serving (excluding unknown items): 354.5 Calories; 14.7g Fat (35.7% calories from fat); 11.9g Protein; 47.9g Carbohydrate; 61mg Cholesterol; 946mg Sodium. Exchanges: 3 Grain(Starch); ½ Non-Fat Milk; 3 Fat.

# English Muffins

Makes 24 servings. Prep Time: 2 hours and 30 minutes

**2 tbsp yeast**
**1 ¾ c warm water, 105-115 degrees**
**2 tsp sugar**
**1 tbsp powdered milk**
**1 tsp salt**
**2 tbsp shortening, softened**
**4 c whole wheat flour**
**½ c cornmeal, for frying**

1. Sprinkle yeast and sugar over water in a large bowl. Let stand 5 minutes, until bubbly.
2. Add salt, shortening and half the flour and dry milk. Beat for 5 minutes.
3. Let sit 1-1 ½ hours, until "sponge" has risen and collapsed. Add remaining flour and knead well.
4. Roll ½-inch thick. Cut with greased rings (tuna cans with ends cut off work well). Let rise in the rings until double in size. If you don't have sufficient rings, the muffins will rise fine without them, only the sides won't have their traditional angular shape. Re-roll dough from cut-outs for more muffins.
5. Bake on a dry griddle sprinkled with cornmeal, over medium heat. Turn once. Griddle may be greased, if necessary.
6. Fork split muffins, if desired, by poking fork tines through center of muffin all the way around the edge and then gently pulling apart. Makes 24 muffins.

*Try soft white wheat or all-purpose flour for a lighter muffin.*

Try these for breakfast or as mini-pizzas.

Per serving (excluding unknown items): 93.7 Calories; 1.6g Fat (14.8% calories from fat); 3.5g Protein; 17.6g Carbohydrate; 0mg Cholesterol; 92mg Sodium. Exchanges: 1 Grain(Starch); ½ Fat.

# Flour Tortillas

Makes 10 servings. Prep Time: 45 minutes

---

**5 c all-purpose flour**
**5 tsp baking powder**
**2 tsp salt**
**¼ c shortening**
**2 ¼ c water**

1. Stir together dry ingredients in a large bowl. Cut in shortening with a pastry blender or two knives until mixture resembles coarse crumbs.
2. Make a well in center and add just enough water to make a dry dough.
3. Form a ball and knead until satiny and smooth.
4. Divide into 10 or 15 portions, depending on desired tortilla size. A good size is 8-10 inches in diameter.
5. Roll out in circles. Cook on a hot, dry griddle, browning one side and turning over to brown other side.
6. Stack in a covered dish to steam or under a dish towel to keep soft.

**Hard Shell Tacos**: Heat one-inch deep oil in a 3-quart saucepan to 375 degrees (will brown a one-inch cube of bread in about 25 seconds). Slide tortilla into oil. Fold in half with tongs or two forks and hold so a one-inch space remains between halves. Fry, turning occasionally until crisp and golden brown, Drain.

**Tortilla Chips**: Cut tortillas into wedges. Fry as for Hard Shell Tacos.

*Soft white wheat flour or whole wheat flour may be substituted for all or part of the flour.*

Serve as burritos, enchiladas, tacos, or "wraps."

---

Per serving (excluding unknown items): 274.0 Calories; 5.7g Fat (19.1% calories from fat); 6.5g Protein; 48.3g Carbohydrate; 0mg Cholesterol; 611mg Sodium. Exchanges: 3 Grain(Starch); 1 Fat.

# Hoe Cakes

Makes 4 servings. Prep Time: 35 minutes

**2 c cornmeal, whole-grain**
**1 tsp salt**
**2 tbsp shortening, melted**
**boiling water**
**2 tbsp shortening**

1. Preheat heavy, oven-proof skillet in 375 degree oven. Mix together cornmeal, salt and shortening. Add enough boiling water to make dough easy to form.
2. Grease preheated skillet with remaining shortening. Form dough into 4 small, flattened loaves and place in skillet.
3. Bake about 25 minutes. Or dough may be cooked slowly on top of the stove, turning when brown.

**Corn Pone**: Increase shortening in dough to 4 tbsp, use ¾ c. boiling water, and add 1 tsp baking powder and about ½ c. milk. Add milk last, enough to make dough of proper consistency. Form and bake or cook, as above.

**Ash Cakes**: Use main recipe. Wrap small cakes in any edible, firm leaves, such as cabbage leaves or greens. Bury cakes in ashes and coals of an open or wood stove fire to bake. Time depends on coals, but may take 20-30 minutes.

*Short on pans? These pioneer cakes were cooked on a greased hoe over an open fire.*

---

Per serving (excluding unknown items): 334.1 Calories; 15.0g Fat (39.4% calories from fat); 5.0g Protein; 46.9g Carbohydrate; 0mg Cholesterol; 554mg Sodium. Exchanges: 3 Grain(Starch); 2 ½ Fat.

# Hoe Cake Pancakes

Makes 6 servings. Prep Time: 30 minutes

**2 c cornmeal**
**1 tsp salt**
**2 tbsp shortening**
**1 c hot water**
**1 c cold water**
**4 tbsp shortening**

1. Mix cornmeal, salt and 2 tbsp shortening. Add 1 c. hot water and stir.
2. Thin with cold water to pouring consistency. Meanwhile, heat 2 tbsp shortening on griddle.
3. Pour small pancakes on hot griddle. Fat must be deep enough to nearly cook pancake through on top before turning for pancakes to hold together.

*This pioneer recipe makes use of what little they had some days.*

Per serving (excluding unknown items): 243.9 Calories; 9.3g Fat (34.6% calories from fat); 3.9g Protein; 35.7g Carbohydrate; 0mg Cholesterol; 359mg Sodium. Exchanges: 2 ½ Grain(Starch); 1 ½ Fat.

# Hush Puppies

Makes 12 servings. Prep Time: 45 minutes

**2 ¼ c cornmeal**
**1 tsp salt**
**2 tbsp onion, finely chopped**
**¾ tsp baking soda**
**1 ½ c milk, reconstituted**
**1 tbsp vinegar**
**oil, for frying**

1. Heat one inch oil in large skillet to 375 degrees (a one-inch cube of bread will brown in about 25 seconds).
2. Meanwhile, stir vinegar into milk to make soured milk.
3. Mix cornmeal, salt, soda and onion. Stir in soured milk.
4. Drop by spoonfuls into hot oil, making about 24 hush puppies.
5. Fry until brown, about 2 minutes.

Per serving (excluding unknown items): 114.3 Calories; 1.5g Fat (11.6% calories from fat); 3.2g Protein; 21.7g Carbohydrate; 4mg Cholesterol; 272mg Sodium. Exchanges: 1 ½ Grain(Starch).

# Indian Pumpkin Bread

Makes 6 servings. Prep Time: 30 minutes

**2 c cornmeal**
**1 c mashed pumpkin**
**water**
**2 tbsp shortening**
**½ tsp salt**
**¼ c sugar**

1. Mix together cornmeal and pumpkin. Add enough water to moisten, and work until dough is easy to handle.
2. Form into flat cakes. Baked on a greased sheet, 375-400 degrees about 20 minutes, until browned.
3. Or, cakes may be fried in shortening on a hot griddle, about 5 minutes per side.

*This is an authentic Native American and pioneer recipe, using Native American ingredients.*

Per serving (excluding unknown items): 220.0 Calories; 5.2g Fat (21.1% calories from fat); 4.4g Protein; 39.0g Carbohydrate; 0mg Cholesterol; 3mg Sodium. Exchanges: 2 ½ Grain(Starch); ½ Vegetable; 1 Fat.

# Skillet Toast

Makes 1 serving. Prep Time: 5 minutes

**2 slices whole wheat bread**
**3 tsp shortening**

1. Heat skillet on medium-high. Spread shortening on both sides of bread.
2. Fry bread until lightly toasted. Use honey or jam if desired.

*This is the way toast was sometimes made during the Depression when butter was scarce. It also works if there is no electric toaster!*

Per serving (excluding unknown items): 319.9 Calories; 16.3g Fat (43.9% calories from fat); 8.1g Protein; 38.7g Carbohydrate; 0mg Cholesterol; 443mg Sodium. Exchanges: 2 ½ Grain(Starch); 3 Fat.

# Irish Soda Scone

Makes 8 servings. Prep Time: 45 minutes

**4 c whole wheat flour**
**4 tsp baking powder**
**1 ½ tsp salt**
**¼ c sugar**
**1 tsp baking soda**
**½ c rolled oats**
**½ c shortening**
**6 tbsp powdered milk**
**2 c water**
**1 c raisins, optional**
**4 tbsp shortening, for frying**

1. Heat oven to 400 degrees. Grease a 10-inch ovenproof skillet or 2 flat cookie sheets.
2. Stir together dry ingredients. Cut in shortening with pastry blender or two knives until it resembles coarse meal.
3. Mix in raisins, if desired. Add water and stir with a fork to make a stiff dough. Knead 10 times.
4. Divide dough in half. Pat out evenly in pan or on cookie sheet in two 10-inch circles. Score almost through dough to mark quarters or eighths, with a floured knife.
5. Bake 30-35 minutes. While still hot, brush with milk and sprinkle with sugar. Makes two 10-inch circles.

**Stovetop Scone**: Preheat skillet and 1 tbsp shortening on stovetop set to medium-low. Pat out scone in pan and cut completely through with knife, as above. Cover and bake for 5 minutes. Remove cover and turn over scone quarters. Add 1 tbsp shortening to pan and cook 5 more minutes or more, until browned and cook through on both sides. Repeat for other scone.

Drizzle honey on warm scones for a real treat.

Per serving (excluding unknown items): 502.2 Calories; 22.3g Fat (38.1% calories from fat); 11.2g Protein; 70.4g Carbohydrate; 6mg Cholesterol; 768mg Sodium. Exchanges: 3 Grain(Starch); 1 ½ Fruit; 4 ½ Fat; ½ Other Carbohydrates.

# Navajo Fry Bread

Makes 4 servings. Prep Time: 45 minutes

**2 c all-purpose flour**
**1 tsp baking powder**
**1 tsp salt**
**1 tbsp shortening**
**½ c water**
**¼ c milk, reconstituted**
**oil, for frying**

1. Stir together dry ingredients. Cut in shortening with a pastry blender.
2. Add liquids and mix well, handling as little as possible.
3. Roll out into 1/8-inch thick circles, 8-10 inches across.
4. Fry in hot oil or shortening, ½-inch deep, until fluffy and browned lightly. Drain on brown paper. Sprinkle with cinnamon and sugar, if desired.

*This is an authentic recipe, from a Navajo reservation about 25 years ago. Try soft white wheat or hard red wheat flour—breads may be heavier, but still delicious.*

Per serving (excluding unknown items): 265.8 Calories; 4.3g Fat (14.9% calories from fat); 7.0g Protein; 48.7g Carbohydrate; 2mg Cholesterol; 633mg Sodium. Exchanges: 3 Grain(Starch); ½ Fat.

# Panned Bread

Makes 9 servings. Prep Time: 1 hour and 15 minutes

**1 lb bread dough**
**2 tbsp shortening**
**4 tbsp sugar**
**1 tsp cinnamon**

1. Use dough for half a loaf of bread, or more if desired.
2. Melt shortening in a large skillet. Heat over medium heat.
3. Cut off about 9 handfuls of dough, patting it in flat rounds.
4. Fry in skillet until brown on bottom. The turn and brown on other side.
5. Drain on brown paper. Sprinkle with sugar and cinnamon if desired.

*This was what happened to bread in the old days if they couldn't wait for it to rise and be baked. It also works if you do not have an oven, or just want a delicious change of pace.*

Per serving (excluding unknown items): 207.7 Calories; 9.7g Fat (41.3% calories from fat); 3.1g Protein; 27.8g Carbohydrate; 0mg Cholesterol; 558mg Sodium. Exchanges: 1 ½ Grain(Starch); ½ Fruit; 2 Fat; ½ Other Carbohydrates.

# Boston Brown Bread

Makes 12 servings. Prep Time: 1 hour and 45 minutes

**2 c milk, reconstituted**
**2 tbsp vinegar**
**1 c all-purpose flour**
**1 c cornmeal**
**1 c whole wheat flour**
**1 c raisins, optional**
**¾ c molasses**
**2 tsp baking soda**
**1 tsp slat**

1. Grease 4 cans, 16oz. size, or a 2-quart tube pan.
2. Stir vinegar into milk to make buttermilk substitute.
3, Beat together all ingredients with an egg beater, about 1 minute.
4. Fill tube pan, or fill cans about ²/₃ full. Cover either tightly with foil.
5. Place cans on rack in Dutch oven or steamer. Pour boiling water around cans to level of rack. Cover. Keep water boiling over low heat until wooden pick inserted in center comes out clean, about 1 ½ hours for cans or 2 ½ hours for tube pan. Add boiling water during steaming, if necessary.
6. Immediately unmold bread. Serve warm with cooked apples or applesauce, with Baked Beans, or as Boston Bean Sandwiches.

**Pressure Cooker Variation**: Cook on rack in pressure cooker according to manufacturer's directions, if available (an 8-quart cooker uses 2 ½ c water. Allow a small stream of steam to escape from vent tube for 30 minutes. Set control at 5 and cook 30 minutes more after control jiggles. Reduce pressure instantly by running cold water over pressure cooker.

**Baked Variation**: Heat oven to 325 degrees. Pour batter into a greased 2-quart casserole. Bake about one hour.

Per serving (excluding unknown items): 230.0 Calories; 1.9g Fat (7.3% calories from fat); 5.2g Protein; 49.8g Carbohydrate; 6mg Cholesterol; 240mg Sodium.
Exchanges: 1 ½ Grain(Starch); ½ Fruit; ½ Fat; 1 Other Carbohydrates.

# Steamed Carrot Pudding

Makes 15 servings. Prep Time: 2 hours and 30 minutes

**1 c whole wheat flour**
**½ tsp cloves**
**½ tsp nutmeg**
**1 tsp cinnamon**
**½ tsp baking soda**
**2 tsp baking powder**
**1/3 c shortening**
**3 tbsp water**
**½ c sugar**
**½ c molasses**
**1 c carrots, grated**
**1 c potatoes, grated**
**1 c apple, grated**
**1 c raisins**
**½ c nuts, optional**

1. Cream together shortening and sugar. Add molasses and cream again.
2. Add remaining ingredients and mix well.
3. Turn into a 2-quart greased mold.
4. Cook as for Boston Brown Bread. Serve hot with vanilla sauce, if desired.

*All the fresh grated vegetable and fruit in this make it moist and almost like a pudding rather than a cake. Very nutritious, too, with plenty of vitamins A and C!*

---

Per serving (excluding unknown items): 202.0 Calories; 7.5g Fat (31.8% calories from fat); 2.6g Protein; 33.7g Carbohydrate; 0mg Cholesterol; 100mg Sodium. Exchanges: ½ Grain(Starch); 1 Fruit; 1 ½ Fat; 1 Other Carbohydrates.

# Steamed Chocolate Pudding

Makes 12 servings. Prep Time: 2 hours and 30 minutes

**5 tbsp shortening**
**6 tbsp sugar**
**1 ½ tbsp soy flour**
**¾ c milk, reconstituted**
**1 ½ c whole wheat flour**
**1 tbsp baking powder**
**¼ tsp salt**
**⅓ c cocoa**

1. Cream shortening and sugar thoroughly with a wooden spoon.
2. Stir together dry ingredients and add alternately with milk to creamed mixture.
3. Turn dough into a greased 1-quart mold and cover with foil.
4. Follow steaming directions for Boston Brown Bread, steaming 2-2 ½ hours until set.

**Pressure Cooker Variation**: Follow general directions in Boston Brown Bread recipe. Use 2 ½ c water in 8-quart cooker. Cook with vent steaming for 30 minutes and 5 lbs of pressure for 30 more minutes.

Serve with hard sauce, if desired.

---

Per serving (excluding unknown items): 140.5 Calories; 6.6g Fat (39.3% calories from fat); 3.3g Protein; 19.6g Carbohydrate; 2mg Cholesterol; 144mg Sodium. Exchanges: 1 Grain(Starch); ½ Fruit; 1 ½ Fat; ½ Other Carbohydrates.

# Steamed Christmas Pudding

Makes 12 servings. Prep Time: 3 hours

**1/3 c shortening**
**2/3 c brown sugar**
**1 c whole wheat flour**
**3 tbsp soy flour**
**½ tsp baking powder**
**½ tsp cinnamon**
**½ tsp allspice**
**½ tsp cloves**
**½ c milk, reconstituted**
**1 ½ tsp almond extract, or rum extract**
**½ c raisins**
**¾ c dried fruit, chopped**
**½ c chopped nuts, optional**

1. Thoroughly cream shortening and sugar with a wooden spoon.
2. Stir together dry ingredients.
3. Alternately add dry ingredients and milk to creamed mixture. Add fruits, nuts and flavoring last.
4. Mix well. Turn into a well-greased 1-quart mold. Cover tightly with foil.
5. Follow directions for steaming in Boston Brown Bread. Steam gently 2 ½ - 3 hours, until set.

**Pressure Cooker Variation**: Follow general directions under Boston Brown Bread, but use 3 c water in an 8-quart pressure cooker. Cook with vent steaming for 45 minutes and 5 lbs of pressure for 1 hour more.

Serve warm as pudding or cold as fruit cake.

---

Per serving (excluding unknown items): 214.2 Calories; 9.9g Fat (39.6% calories from fat); 3.5g Protein; 30.5g Carbohydrate; 1mg Cholesterol; 28mg Sodium. Exchanges: ½ Grain(Starch); ½ Fruit; 2 Fat; 1 Other Carbohydrates.

## Steamed Dinner Loaf

Makes 12 servings. Prep Time: 2 hours

**3 c whole wheat flour**
**3 tbsp soy flour**
**½ tsp salt**
**2 tsp baking powder**
**1/3 c powdered milk**
**2 c water**
**1/3 c oil**

1. Stir together dry ingredients. Make a well in center.
2. Pour in liquids and stir until smooth.
3. Pour into a greased 2-quart tube mold or 3 16-oz. cans.
4. Cook as for Boston Brown Bread. Herbs, spices or flavorings may be added as desired. Choose from such as:onion, dill, parsley, chili powder, oregano, basil and garlic, etc.

*This may be used as a tasty alternative to regular bread when you have no yeast and no oven. It slices neatly and may be used for toast or sandwiches. And it requires no kneading!*

---

Per serving (excluding unknown items): 179.0 Calories; 7.8g Fat (37.6% calories from fat); 5.5g Protein; 23.8g Carbohydrate; 3mg Cholesterol; 165mg Sodium. Exchanges: 1 ½ Grain(Starch); 1 ½ Fat.

## Steamed Nut or Oat Bread

Makes 12 servings. Prep Time: 2 hours

**½ c sugar**
**2 tbsp oil**
**1 ½ tbsp soy flour**
**3 tbsp water**
**1 c milk, reconstituted**
**2 ½ c whole wheat flour**
**2 tsp baking powder**
**½ tsp salt**
**1 c chopped nuts, or rolled oats**

1. Whisk together oil, soy flour, water and sugar until light.
2. Alternately add milk and dry ingredients.
3. Cook as for Boston Brown Bread.

---

Per serving (excluding unknown items): 225.8 Calories; 10.2g Fat (38.6% calories from fat); 6.3g Protein; 30.4g Carbohydrate; 3mg Cholesterol; 162mg Sodium. Exchanges: 1 ½ Grain(Starch); ½ Lean Meat; ½ Fruit; 2 Fat; ½ Other Carbohydrates.

## Steamed Holiday Pudding

Makes 12 servings. Prep Time: 2 hours

**1 c flour**
**¼ tsp salt**
**1 tsp baking soda**
**1 tsp nutmeg**
**1 tsp cinnamon**
**1 c bread crumbs**
**1 ¼ c milk, reconstituted**
**$^{1}/_{3}$ c oil**
**¾ c sugar**
**3 tbsp soy flour**
**1 c raisins**
**½ c nuts, optional**

1. Stir together flour, salt, baking soda, soy flour and spices.
2. Soak bread crumbs in milk.
3. Combine oil and sugar and add to bread crumbs.
4. Add dry ingredients, raisins and nuts to bread crumb mixture.
5. Fill individual pudding molds or 16 oz. cans $^{2}/_{3}$ full.
6. Cook as for Boston Brown Bread.

---

Per serving (excluding unknown items): 272.1 Calories; 11.3g Fat (36.4% calories from fat); 4.8g Protein; 39.8g Carbohydrate; 3mg Cholesterol; 242mg Sodium. Exchanges: 1 Grain(Starch); 1 ½ Fruit; 2 Fat; 1 Other Carbohydrates.

## Steamed Pumpkin Pudding

Makes 12 servings. Prep Time: 2 hours

**3 c flour**
**4 tsp baking powder**
**½ tsp salt**
**$^{2}/_{3}$ c shortening**
**1 ½ cans sugar**
**4 ½ tbsp soy flour**
**1 tsp vanilla**
**1 c mashed pumpkin**
**1 ½ c milk, reconstituted**
**1 c chopped nuts, optional**

1. Cream shortening and sugar.
2. Add vanilla and pumpkin, mixing well.
3. Add dry ingredients alternately with milk. Stir in nuts.
4. Pour into 3 one-quart molds, 3 28-oz. size cans or 5 16-oz. size cans.
5. Cook as for Boston Brown Bread.

---

Per serving (excluding unknown items): 420.4 Calories; 19.9g Fat (41.6% calories from fat); 7.1g Protein; 55.6g Carbohydrate; 4mg Cholesterol; 228mg Sodium. Exchanges: 2 Grain(Starch); ½ Lean Meat; ½ Vegetable; 1 ½ Fruit; 3 ½ Fat; 1 ½ Other Carbohydrates.

# BAKED BREADS

## Yeast Breads and Quick Breads

All breads need some leavening agent to make air bubbles within the dough to lighten it. In yeast bread, the air bubbles are produced by yeast, actually a plant, growing and digesting sugar in the dough. Carbon dioxide gas, a by-product of their digestion, produces the leavening. Generally, these breads also must be kneaded to develop the strands of gluten, a protein in wheat flour, which gives yeast breads their stretchy, flexible texture. This entire process takes some time.

On the other hand, quick breads are, literally, quick. Leavening is produced by a chemical process, by either baking powder alone or a combination of baking soda and some acidic source, such as cream of tartar, vinegar, sour milk, applesauce, etc. Usually there is not much gluten development or kneading and these breads tend to have a more crumbly texture, like a cake.

## Using Whole Wheat Flour

All-purpose flour is made from the same type of grain as whole wheat flour, but the bran and germ portions of the grain are removed. These portions are actually very nutritious, but they do present some challenges when converting standard recipes to whole wheat flour.

Firstly, the flour tends to be slightly coarser and the flakes of bran and germ tend to settle out. The coarser texture of the flour will cause coarser textured breads.

Secondly, the coarser textured flour and naturally waterproof bran takes longer to absorb moisture. In quick breads, this means that the dough will thicken if it sits before cooking, and this may improve the quality of the final product. On the other hand, you may look at the dough immediately after mixing and decide it looks to thin, add extra flour to thicken it, only to end up with a very dry product after baking because you added too much flour for the amount of liquid. In the case of pancakes, you may mix the batter so that the first pouring on the griddle is just the right consistency, only to find that by the second pouring is too thick. You add a little water, get it right again and make the second batch on the griddle. By the third pouring, it is too thick AGAIN, and you add a little more water. Usually, at least at our house, the subsequent pourings do not require more thinning because the flour has finally absorbed all the liquid it can absorb.

In yeast breads, this slow water absorption means that you should make a "sponge" dough first, by adding half the flour and beating hard to develop gluten and help the flour absorb as much moisture as possible. After the beating, the sponge is allowed to sit about 15 minutes to continue moisture absorption as well as a first rising of the yeast. Then the remaining flour is stirred and kneaded in. This prevents adding too much flour at the start and ending up with a very dry, tough loaf of bread.

Thirdly, there are some uses of all-purpose flour in general cooking which are decidedly unsuitable for whole wheat flour, particularly the high-protein, hard red type. These include thickening, biscuits, cakes, pies and pastries. These items require more starch, less protein and a lighter texture than is afforded by high-protein whole wheat flour. For these purposes, low protein flour substitutes will provide results more like those we are accustomed to. Arrowroot starch or cornstarch may be used instead of flour for thickening gravies, etc. Use about half the amount of flour called for. Ground soft white wheat, which is higher in starch and lower in protein, makes a lighter flour that will produce better results for biscuits, cakes and pastries.

# Baking Powder Biscuits

Makes 6 servings. Prep Time: 25 minutes

**1/3 c shortening**
**1 ¾ c all-purpose flour**
**2 ½ tsp baking powder**
**¾ tsp salt**
**3 tbsp powdered milk**
**¾ c water**

1. Heat oven to 450 degrees. Stir together dry ingredients.
2. Cut in shortening using a pastry blender or two knives, until mixture looks like fine crumbs.
3. Stir in just enough water so dough leaves sides of bowl and rounds up into a ball.
4. Turn out onto a lightly floured surface. Knead lightly about 12 times.
5. Roll ½ inch thick. Cut into circles, squares or strips. Place on ungreased baking sheet.
6. Bake until golden brown, 10-12 minutes. Remove immediately from pan. Makes about 12 biscuits.

**Stir-N-Roll Biscuits**: Use oil instead of shortening. Stir in oil with water (no need for pastry blender). Roll dough between sheets of waxed paper on a dampened surface (to keep it from slipping), or drop by spoonfuls onto cookie sheet.

**Drop Biscuits**: Use regular recipe, but increase water to 1 c. Drop dough by spoonfuls onto cookie sheet. This recipe works better for heavier flour, such as the hard red whole wheat, or even the soft white wheat flour. It will make a lighter biscuit that will rise more.

**Pumpkin Biscuits**: Add 1 cup cooked, mashed pumpkin. Adjust flour if needed. Add sugar and cinnamon to taste.

*This is one of those recipes in which the switch to whole wheat flour makes a big difference, but feel free to use it if it is all that you have!*

---

Per serving (excluding unknown items): 254.2 Calories; 12.8g Fat (45.4% calories from fat); 4.8g Protein; 29.8g Carbohydrate; 4mg Cholesterol; 434mg Sodium. Exchanges: 2 Grain(Starch); 2 ½ Fat.

## Applesauce Raisin Bread

Makes 10 servings. Prep Time: 1 hour and 10 minutes

**1 ½ c whole wheat flour**
**1 ½ tsp baking powder**
**1 tsp salt**
**½ tsp baking soda**
**1 ½ tsp cinnamon**
**1 c rolled oats**
**3 tbsp soy flour**
**¾ c brown sugar**
**¾ c raisins**
**1 ¼ c applesauce**
**1/3 c oil**
**½ c milk, reconstituted**

1. Stir together all dry ingredients.
2. Add applesauce, oil and milk. Stir only until ingredients are moistened.
3. Grease and flour loaf pan, 9x5x3 inches. Pour batter into pan.
4. Bake at 350 degrees 55-60 minutes. Remove from pan onto cooling rack immediately and cool before serving. Makes one loaf.

Per serving (excluding unknown items): 290.9 Calories; 9.0g Fat (26.4% calories from fat); 5.1g Protein; 51.1g Carbohydrate; 2mg Cholesterol; 347mg Sodium. Exchanges: 1 ½ Grain(Starch); 1 Fruit; 1 ½ Fat; 1 Other Carbohydrates.

## Whole Wheat Quick Bread

Makes 8 servings. Prep Time: 1 hour and 10 minutes

**3 tbsp shortening**
**3 tbsp sugar**
**1 ¼ c milk, sour**
**1 tsp baking soda**
**1 c whole wheat flour**
**1 c all-purpose flour**
**½ tsp salt**
**1 ½ tbsp soy flour**

1. Sour milk by combining with 1 tbsp vinegar. Let stand about 5 minutes.
2. Mix together all ingredients and bake in a well-greased loaf pan.
3. Bake at 350 degrees for about 1 hour. Turn out and cool on rack.

*This Depression-era recipe (modified to use soy as an egg replacement) uses baking soda instead of yeast as a leavening agent.*

Per serving (excluding unknown items): 196.1 Calories; 6.7g Fat (30.2% calories from fat); 5.3g Protein; 29.6g Carbohydrate; 5mg Cholesterol; 310mg Sodium. Exchanges: 1 ½ Grain(Starch); ½ Fruit; 1 ½ Fat; ½ Other Carbohydrates.

# Bean-Zucchini Bread

Makes 12 servings. Prep Time: 1 hour

**1 c pinto beans, cooked, pureed**
**4 ½ tbsp soy flour**
**6 tbsp water**
**1 ½ c sugar**
**1 c oil**
**1 ½ c zucchini, shredded**
**1 tsp vanilla**
**2 c whole wheat flour**
**1 tsp salt**
**1 tsp baking soda**
**2 ½ tsp baking powder**
**2 tsp cinnamon**

1. Preheat oven to 350 degrees. Grease two 9x5x3 loaf pans.
2. Using 1 c bean puree, blend together puree, soy flour, water, oil and sugar.
3. Add zucchini and vanilla.
4. Stir in dry ingredients until blended. Pour into greased pans.
5. Bake 40-50 minutes, until wooden pick inserted comes out clean. Remove from pan and cool on rack. Makes 2 large loaves.

**Carrot-Bean Bread**: Substitute 1 c shredded carrot for the zucchini.

*The beans, combined with the whole wheat flour, forms complementary proteins and make this sweet bread especially nutritious.*

Per serving (excluding unknown items): 358.2 Calories; 19.0g Fat (46.3% calories from fat); 4.8g Protein; 45.0g Carbohydrate; 0mg Cholesterol; 361mg Sodium. Exchanges: 1 Grain(Starch); 1 ½ Fruit; 4 Fat; 1 ½ Other Carbohydrates.

# Zucchini Bread

Makes 12 servings. Prep Time: 1 hour and 20 minutes

**2/3 c shortening**
**1 ½ c sugar**
**3 1/3 c whole wheat flour**
**6 tbsp soy flour**
**2 tsp baking soda**
**½ tsp baking powder**
**1 tsp cinnamon**
**1 tsp cloves**
**2/3 c raisins**
**2 tsp vanilla**
**3 c zucchini, shredded**
**1 c water**

1. Heat oven to 350 degrees. Grease two loaf pans, 9x5x3 or three pans 8 ½ x 4 ½ x 2 ½.
2. cream together shortening and sugar. Add zucchini and water.
3. Add all other ingredients and mix thoroughly.
4. Pour into pans. Bake until wooden pick inserted in center comes out clean, 60-70 minutes.
5. Cool slightly before removing from pans. Makes 2 large or three medium loaves.

**Pumpkin Bread**: Omit vanilla. Replace zucchini with 2 c mashed pumpkin.

**Carrot Bread**: Replace zucchini with 3 c shredded carrots.

*The pumpkin or carrot versions make this bread a good source of vitamin A.*

---

Per serving (excluding unknown items): 354.1 Calories; 12.7g Fat (30.8% calories from fat); 6.1g Protein; 57.9g Carbohydrate; 0mg Cholesterol; 230mg Sodium. Exchanges: 1 ½ Grain(Starch); 2 Fruit; 2 ½ Fat; 1 ½ Other Carbohydrates.

# Basic Muffins

Makes 6 servings. Prep Time: 30 minutes

**2 c whole wheat flour**
**$^1/_3$ c sugar**
**1 tbsp baking powder**
**3 tbsp powdered milk**
**1 ½ tbsp soy flour**
**1 c water**
**$^1/_3$ c oil**

1. Heat oven to 400 degrees. Grease or line 12 small or 6 large muffin cups.
2. Stir together dry ingredients. Make a well in center.
3. Pour in water and oil, and stir only until moistened (batter will still be lumpy).
4. Fill cups ¾ full. Bake until golden brown, 20-25 minutes. Remove immediately from pan. Makes 12 small or 6 large muffins. These traditional muffins will be bread-like and have cauliflower-like tops. They will not be like today's very sweet and cake-like muffins.

**Apple Muffins**: Stir in 1 chopped apple, ½ tsp cinnamon

**Blueberry Muffins**: Stir in 1 c. fresh or ¾ c. canned blueberries.

**Oatmeal-Raisin** : Stir in 1 c. raisins, ½ tsp nutmeg, 1 tsp cinnamon. Decrease flour to 1 c. Add 1 c oats.

**Pumpkin Muffins**: Stir in ½ c. pumpkin, ½ c. raisins, 1 tsp cinnamon, ½ tsp nutmeg, ½ tsp cloves. Add more flour if needed.

**Chocolate Muffins**: Increase sugar to 1 ½ c and soy flour to 3 tbsp. Add ¾ c cocoa with dry ingredients. Add about 1 ½ c more water. If desired, stir in 1 c chocolate chips after wet ingredients are added.

Per serving (excluding unknown items): 312.4 Calories; 14.2g Fat (39.1% calories from fat); 7.0g Protein; 42.7g Carbohydrate; 4mg Cholesterol; 200mg Sodium. Exchanges: 2 Grain(Starch); ½ Fruit; 3 Fat; ½ Other Carbohydrates.

## Gingerbread Muffins

Makes 6 servings. Prep Time: 30 minutes

**2 ½ c whole wheat flour**
**½ c sugar**
**2 ½ tsp baking powder**
**1 tsp baking soda**
**1 tsp cinnamon**
**1 tsp ginger**
**¼ tsp nutmeg**
**1 ½ tbsp soy flour**
**3 tbsp powdered milk**
**1 1/8 c water**
**1 c molasses**
**1/3 c melted shortening**

1. Preheat oven to 400 degrees. Mix and bake according to Basic Muffin recipe. Makes 12 small muffins.

*The molasses makes these muffins moist and as dark as chocolate.*

Per serving (excluding unknown items): 408.5 Calories; 2.4g Fat (5.0% calories from fat); 8.4g Protein; 93.6g Carbohydrate; 4mg Cholesterol; 401mg Sodium. Exchanges: 2 ½ Grain(Starch); 1 Fruit; ½ Fat; 3 ½ Other Carbohydrates.

## Soy Muffins

Makes 6 servings. Prep Time: 30 minutes

**1 c whole wheat flour**
**1 c soy flour**
**1 tbsp baking powder**
**2 tbsp sugar**
**1 tbsp shortening, melted**
**¼ c powdered milk**
**1 c water**

1. Preheat oven to 400 degrees. Mix and bake according to Basic Muffin recipe. Makes 12 small muffins.

*The soy flour, combined with the wheat in this recipe, forms complementary proteins. Soy flour also adds a special sweetness, moistness and nut-like flavor.*

Per serving (excluding unknown items): 191.5 Calories; 6.8g Fat (30.4% calories from fat); 9.0g Protein; 26.2g Carbohydrate; 5mg Cholesterol; 205mg Sodium. Exchanges: 1 ½ Grain(Starch); ½ Lean Meat; ½ Fruit; 1 Fat; ½ Other Carbohydrates.

# Corn Bread

Makes 9 servings. Prep Time: 40 minutes

**1 c cornmeal**
**1 c whole wheat flour**
**2 tsp baking powder**
**2 tsp sugar**
**1 tsp salt**
**½ tsp baking soda**
**¼ c shortening**
**¼ c powdered milk**
**3 tbsp soy flour**
**1 ½ tbsp vinegar**
**1 ¾ c water**

1. Heat oven to 450 degrees. Mix all ingredients and beat well, 30 seconds.
2. Pour into a greased 9-inch round layer or 8-inch square pan. Bake 25-30 minutes, until golden brown. Or preheat cast iron cornbread pan (any shape mold) in oven. Brush with oil or shortening. Pour in batter and bake about 20 minutes, depending on size of individual molds. Smaller molds cook faster Makes 1 pan or 8-12 individual cornbreads.

**Cornbread Muffins**: Fill 12-14 greased or lined c. Bake 20 minutes.

**Chili Cornbread**: Stir in ½ c green chilies, drained, and ½ tsp chili powder.

**Double Cornbread**: Stir in 1 c while kernel corn.

**Skillet Cornbread**: Preheat cast iron 10-inch skillet (or other ovenproof skillet) in oven. Grease generously. Pour in batter and bake 20 minutes. Or bake on top of the stove, as in Irish Soda Scone recipe.

*Using the preheated skillet or mold make a delicious, crispy outside crust and a more moist interior on the cornbread.*

---

Per serving (excluding unknown items): 181.4 Calories; 7.5g Fat (36.3% calories from fat); 4.7g Protein; 24.9g Carbohydrate; 3mg Cholesterol; 403mg Sodium. Exchanges: 1 ½ Grain(Starch); 1 ½ Fat.

# Pioneer Johnnycake

Makes 9 servings. Prep Time: 40 minutes

---

**1 ½ c cornmeal**
**½ c whole wheat flour**
**1 tsp baking soda**
**½ tsp salt**
**1 ½ tbsp soy flour**
**4 tbsp powdered milk**
**1 ½ c water**
**1 ½ tbsp vinegar**
**1 ½ tbsp molasses**

1. Mix together dry ingredients.
2. Slowly stir in water, molasses and vinegar.
3. Beat hard 2 minutes.
4. Pour into a greased 9-inch square pan. Bake 25-30 minutes at 425 degrees in a preheated oven.
5. Cut into 9 squares when mostly cool. Makes one 9x9 pan.

*This recipe was adapted (soy flour substituted for egg) from an authentic pioneer recipe. Note that it contains only molasses for sweetener and no fat of any kind. These items were in short supply on the prairies!*

---

Per serving (excluding unknown items): 137.7 Calories; 1.6g Fat (10.6% calories from fat); 4.1g Protein; 26.9g Carbohydrate; 3mg Cholesterol; 275mg Sodium. Exchanges: 1 ½ Grain(Starch).

# Whole Wheat Crackers

Makes 10 servings. Prep Time: 30 minutes

**1/3 c shortening**
**1/3 c milk, reconstituted**
**2 tbsp honey**
**1 c whole wheat flour**
**2/3 c all-purpose flour**
**2 tbsp yellow cornmeal**
**½ tsp baking powder**
**¼ tsp salt**

1. Heat oven to 350 degrees. Mix shortening and honey in 2-quart bowl. Stir in milk.
2. Stir in dry ingredients until well-blended.
3. Divide dough in half. Roll each half on greased cookie sheet into a 10x15 inch rectangle. Place cookie sheet on a damp, thin towel to prevent slipping.
4. Cut dough into 1 ½-inch squares. Bake until light brown, 10-12 minutes. Repeat with other half. Makes 10 dozen crackers.

*Soft white wheat flour is a good substitute for all-purpose flour in this recipe. If only hard red wheat flour is available, that also may be substituted, but results will be a little coarser and more crumbly.*

Per serving (excluding unknown items): 155.7 Calories; 7.4g Fat (41.8% calories from fat); 2.9g Protein; 20.3g Carbohydrate; 1mg Cholesterol; 76mg Sodium. Exchanges: 1 Grain(Starch); 1 ½ Fat.

# Basic Whole Wheat Bread

Makes 18 servings. Prep Time: 2 hours and 15 minutes

**2 tbsp yeast**
**2 ¾ c warm water, 105-115 degrees**
**3 tbsp honey, or sugar**
**3 tbsp oil**
**1 tbsp salt**
**$^{1}/_{3}$ c powdered milk**
**½ c potato flakes**
**7 c whole wheat flour**

1. In a large mixing bowl, sprinkle yeast on water (if no thermometer, test water on wrist-should be slightly warmer than "hot tub" temperature.) Pour in oil and honey. DO NOT STIR. Let stand 5 minutes, until bubbly.
2. Proceed only if yeast is bubbly. If not, yeast is probably dead and will not rise bread. Add salt, milk, potato flakes and half the flour. Stir hard for about 5 minutes to develop gluten.
3. Let this "sponge" stand for about 15 minutes to rise.
4. Stir down. Add remaining flour, adjusting the amount to form a dough that is just thick enough to knead by hand. Knead 5 minutes on a smooth, oiled surface.
5. Form into 2 loaves, by patting into a large rectangle, rolling up tightly along the short edge, pinching the ends and tucking them under the loaf. Put loaves into greased 8 ½ x 4 ½ inch loaf pans. Bread may not rise well in a larger pan.
6. Let rise until slightly above top of pan. You may do the rising in a slightly warm (under 170 degrees) oven to speed up the process. Rising will take 30-60 minutes.
7. Bake 20 minutes at 375 degrees and 20 minutes more at 350 degrees. Cover with foil during baking of loaves brown too much, or lower oven temperatures 25 degrees next time.
8. Immediately turn out onto cooling rack and over with cloth or foil loosely until cool to keep crust soft. Will keep at least 5 days in a plastic bag on a shelf. Makes 2 loaves.

**100% Whole Wheat Bread**: Use ¾ c whole wheat flour in place of potato flakes and milk, if these are unavailable. These lighten the loaf, provide better texture, better browning and more protein.

**Soy-Enriched Bread**: Use 6 ¼ c whole wheat flour and ¾ c soy flour instead of the 7 c flour. This increases the protein, moistness and gives a sweet, nutty flavor. But it may also make a slightly denser loaf.

**Cinnamon Swirl Bread**: After patting out loaf, sprinkle with 1 T water, ¼ c sugar and 1 tsp cinnamon. Roll up tightly to form loaf.

**Oat Bread**: Decrease whole wheat flour to 6 c and add 1 c rolled oats. Add the oats with the first half of the flour to make the "sponge."

**Raisin Bread**: Before starting bread, heat 1 ½ c raisins in enough water to cover until plump, about 5 minutes. Drain and use water to begin bread. Add plump raisins with first half of flour.

**Cracked Wheat, Oat Bran, 7-Grain, Etc Bread**: Any dry, cracked or crushed grains may be substituted for part of the whole wheat flour. Never use more than ½-1 c in two loaves, or bread may be dense, dry and crumbly. Always add with the first half of the flour.

**Rye Bread**: Decrease whole wheat flour to 5 ½ c Use 1 ½ c ground rye. 1 T caraway seed may be added.

*A cleaning tip: If oil is measured into liquid measuring cup first, then honey on top (it then sinks to the bottom), the oil should keep the honey from sticking in the measuring cup.*

---

Per serving (excluding unknown items): 209.1 Calories; 3.8g Fat (15.5% calories from fat); 7.6g Protein; 39.2g Carbohydrate; 2mg Cholesterol; 370mg Sodium. Exchanges: 2 ½ Grain(Starch); 1 Fat.

# Basic Whole Wheat Bread:

# Theme and Variations

Makes 18 servings. Prep Time: 2 hours and 20 minutes

**2 lb bread dough**

Any of the Basic Whole Wheat breads can be varied by forming into other shapes, as listed below.

**French Bread**: Using dough for 1 loaf, roll dough into long, narrow rectangle. Roll along long edge to form French bread shape. Cut 5 diagonal gashes, ¼-inch deep in top of loaf using sharp knife. Brush with cold water and sprinkle with sesame seed if desired. Bake 30-35 minutes in 375 degree oven. Makes one loaf.

**Pan Rolls**: Using dough for one loaf, break off small pieces of dough, about 2 ounces or ¼ c each, and form into small balls. Place in greased casserole dish or 9x13 pan, about ¼ inch apart, allowing room for rising. Let rise before baking. Bake 375 degrees, 25-30 minutes. Makes 16 rolls.

**Hamburger Buns**: Using dough for one loaf, break off pieces of dough the size of a large egg. Form ball and then flatten. Place on greased cookie sheet, leaving plenty of room for rising. Bake 375 degrees, 20-25 minutes. Make 10 buns.

**Parker House Rolls**: Using dough for one loaf, roll dough ½-inch thick. Cut with biscuit cutter or glass. Dip in oil. Fold in half and press edges of round together. Place about ½ inch apart on ungreased cookie sheet. Bake, after rising, in 375 degree oven about 20 minutes. Makes about 16 rolls.

**Cloverleaf Rolls**: Using dough for one loaf, form dough into 1-inch balls. Place 3 balls in each greased muffin c. Brush with oil. Bake in 375 degree oven about 20 minutes. Make about 12 rolls.

**Crescent Rolls**: Use dough for 2 loaves, but divide into 3 portions. Roll each third into an 8-inch circle. Cut each into12 wedge-shaped pieces like a pie. Starting at wide end, roll each wedge into a crescent or butterhorn shape. Place on greased cookie sheet, 2-inches apart. Makes a total of 36 rolls. Let rise. Bake 375 degrees, 20 minutes.

**Cinnamon Rolls**: Use dough for one loaf. Roll into long, narrow rectangle, about 9x15 or 9x18 inches. Spread with 2 tbsp shortening. Sprinkle with ½ c sugar (white or brown), 2 tsp cinnamon and ½ c each raisins and nuts (opt) or 1 c chopped apples. Beginning with long side, roll up dough and seal edge by pinching with fingers. To "cut" into rolls, take a string or dental floss about 18 inches long. With one end in each hand, slip string under roll to desired cutting place. Bring ends of string up and cross as if tying knot. Pull until string cuts completely through roll without spoiling its shape. Place in greased 9x13 pan, fairly close together, allowing some space for rising, but so edge will be together when dough has risen. Bake at 375 degrees for 25-30 minutes. Drizzle with glaze , if desired, while still slightly warm. See Swedish Tea Ring recipe. Makes about 10 rolls.

**Sticky Buns**: Form as for cinnamon rolls. Stir together 1 c brown sugar and 12 c water in bottom of greased 9x13 pan. Then place rolls in pan. After baking, immediately turn out onto serving plate. Syrup will drizzle down and into rolls. Makes about 10 rolls.

**Swedish Tea Ring**: Use dough for 2 loaves, but make a thicker, slightly larger dough rectangle. Form as for cinnamon rolls, but do not cut into slices. Instead, place on greased baking sheet, joining ends to form a ring. Seal together well. With scissors, make cuts into ring about ²/₃ of the way through and one inch apart, all the way around the ring. Turn each section on its side, fanning out the sections. Let rise and bake at 350 degrees, 25-30 minutes. Drizzle with glaze:1 c powdered sugar, ½ tsp vanilla and 1 T water. Decorate with red cherries and greenery for Christmas.

# Batter Bread

Makes 8 servings. Prep Time: 1 hour and 45 minutes

**1 tbsp yeast**
**1 ¼ c warm water, 105-115 degrees**
**1 ½ tbsp sugar**
**1 ½ tbsp powdered milk**
**¼ c oil**
**3 tbsp soy flour**
**1 tsp salt**
**3 c whole wheat flour**

1. Sprinkle yeast and sugar over warm water in a large mixing bowl. Let sit 5 minutes, until bubbly.
2. Add all ingredients, but half the flour. Stir together until blended. Then beat thoroughly for 2 minutes.
3. Stir in remaining flour until smooth.
4. Cover and let rise until double, about 30 minutes. This rising is optional, but it produces a better texture.
5. Stir down batter by beating 25 strokes. Spread in a greased 2-quart casserole and let rise again, about 40 minutes.
6. Bake at 375 degrees until loaf is brown and sounds hollow when tapped, 40-45 minutes.
7. Immediately remove from pan. Mexican, Italian or onion-dill seasonings may be added at step 2, for variety. Makes 1 casserole pan.

*This is a yeast bread that requires no kneading! The dough contains more water, so it is softer and is more of a batter than a dough. By slicing the casserole bread through the middle one direction and in thin slices the other direction, you can make bread slices for toasting or sandwiches.*

Per serving (excluding unknown items): 242.3 Calories; 8.5g Fat (30.1% calories from fat); 7.8g Protein; 36.8g Carbohydrate; 1mg Cholesterol; 276mg Sodium. Exchanges: 2 ½ Grain(Starch); 1 ½ Fat.

# Easy Soft Pretzels

Makes 12 servings. Prep Time: 45 minutes

---

**1 tbsp yeast**
**1 ¾ c warm water, 105-115 degrees**
**3 ½ c whole wheat flour**
**1 tbsp sugar**
**½ tsp salt**
**¼ c milk, reconstituted**
**2 tbsp coarse salt, if available**

1. Dissolve yeast and sugar in warm water, as for Basic Whole Wheat Bread. Let sit 5 minutes, until bubbly.
2. Add half the flour, ½ tsp salt. Stir 3 minutes, briskly, to develop gluten.
3. Stir in enough remaining dough, or add up to ½ c more, to make dough easy to handle.
4. Knead on a lightly oiled surface 5 minutes, until smooth and elastic.
5. Cover and let rise in a warm place until double, about one hour. Punch down dough. (This step optional).
6. Heat oven to 425 degrees. Divide dough in half. Cut each half into 6 equal pieces. Roll each piece into a 15-inch rope. Place rope on greased cookie sheet.
7. To form pretzel, bring left end of rope over to the middle to form a loop. Bring right end of rope up and over the first loop, towards the middle, to make pretzel shape.
8. Brush with milk or reconstituted egg. Sprinkle with coarse salt.
9. Bake 15-20 minutes, until brown. Cool on wire rack. Makes 12 pretzels.

Serve with mustard, if desired.

---

Per serving (excluding unknown items): 128.8 Calories; 0.9g Fat (5.7% calories from fat); 5.3g Protein; 27.1g Carbohydrate; 1mg Cholesterol; 1035mg Sodium. Exchanges: 1 ½ Grain(Starch).

# Peach Focaccia

Makes 6 servings. Prep Time: 2 hours

---

**1 lb whole wheat bread dough**
**2 tbsp shortening, melted**
**3 c peaches, thinly sliced**
**3 tbsp sugar**
**¼ tsp nutmeg**
**1 tsp cinnamon**

1. Use dough for one-half loaf of bread from Basic Whole Wheat Bread recipe. Grease 10x15 inch baking pan. Place dough in pan and stretch evenly to fill pan. Cover lightly with towel or plastic wrap and let rise about 45 minutes. It should be puffy.
2. Brush dough with 1 tbsp shortening. Peaches may be fresh or canned, and any soft fruit may be substituted. Apples and raisins may also be used. Arrange peach slices on dough and brush with remaining shortening.
3. Mix sugar and nutmeg (allspice may be substituted) and cinnamon. Sprinkle over fruit.
4. Bake in 350 degree oven until well browned on edges and bottom, about 40 minutes. Serve warm. Makes one 10x15 inch pan.

*This is like a sweet, fruit pizza! Delightful for a special breakfast or an unusual dessert.*

May be stored overnight at room temperature before serving.

---

Per serving (excluding unknown items): 100.0 Calories; 4.4g Fat (37.3% calories from fat); 0.6g Protein; 16.0g Carbohydrate; 0mg Cholesterol; 0mg Sodium. Exchanges: 1 Fruit; 1 Fat; ½ Other Carbohydrates.

# Pinto Bread

Makes 9 servings. Prep Time: 2 hours and 30 minutes

**1 c bean puree**
**1 c warm water, 105-115 degrees**
**1 tbsp honey**
**1 tbsp yeast**
**2 tbsp oil**
**1 ½ tsp salt**
**3 c whole wheat flour**

1. Follow instructions for Basic Whole Wheat Bread. Sugar may be substituted for honey.
2. Add the bean puree with the first half of the flour.
3. Add more or less flour, to make dough of good consistency. Bake as for Basic Whole Wheat Bread. Makes one loaf.
4. This bread will be more dense and moist than the Basic Bread. By combining the wheat grain and the beans, it provides complementary proteins.

Per serving (excluding unknown items): 226.5 Calories; 3.9g Fat (14.7% calories from fat); 7.4g Protein; 43.6g Carbohydrate; 0mg Cholesterol; 386mg Sodium. Exchanges: 2 ½ Grain(Starch); 1 Fat.

# Pita Bread (Pocket Bread)

Makes 8 servings. Prep Time: 2 hours

---

**1 c warm water, 105-115 degrees**
**2 tbsp yeast**
**1 tbsp sugar**
**1 tsp oil**
**1 tsp salt**
**2 ¾ c whole wheat flour**

1. This recipe will produce better results if half whole wheat and half white or all-purpose flour are used. Follow instructions for Basic Whole Wheat Bread dough, up to the point of forming loaves.
2. Place dough in a greased bowl; turn to grease top. Cover and let rise in a warm place, such as inside a cool or under 170 degree oven, about an hour or until doubled.
3. Turn out onto a lightly oiled surface. Cut into 8 equal pieces. Roll each in a ball, then flatten with fingers to a 4-inch round. Let rest 5-10 minutes.
4. Roll each into a 6-inch round, about ¼ inch thick. Place on oiled or floured waxed paper or board. Cover with waxed paper and a light towel.
5. Let rise as before, about 30 minutes or until rounds look smooth and slightly raised.
6. Meanwhile, place clean, unglazed 6-inch red clay tiles or a large, heavy baking sheet in middle of lowest oven rack, leaving at least 3 inches from oven walls.
7. Preheat oven to 500 degrees. After about 20 minutes of preheating, carefully transfer risen rounds, floured side down with wide spatula or wide piece of cardboard, one at a time, to hot tiles of baking sheet.
8. Bake 4-5 minutes, or until rounds balloon up and are very lightly browned.
9. Remove from oven; place on waxed paper; cover with waxed paper and a towel, so their own steam will keep them soft.
10. Cool, then press down to flatten. To serve, slit at edge and open to make one large pocket, or cut in half and pull open to make two pockets. Makes 8.

**Flat Pitas**: If tiles or baking sheet are not preheated, or if oven temperature is not high enough (or for a variety of other reasons), pitas may not "balloon" as desired. They are still delicious and may be used in a variety of ways: individual pizzas, instead of tortillas for "wraps," or cut up and toasted for "chips."

---

Per serving (excluding unknown items): 159.8 Calories; 1.5g Fat (7.8% calories from fat); 6.8g Protein; 32.6g Carbohydrate; 0mg Cholesterol; 271mg Sodium. Exchanges: 2 Grain(Starch); ½ Fat.

# POPCORN SURPRISES

In an age of microwave popcorn and electric air poppers, it seems important to include a standard popcorn recipe. Also, as popcorn is a good starch food that can cook up in a hurry, not requiring a grinder or a lot of fuel, additional untraditional recipes are included here to use popcorn in our daily meals as well as snacks. Check out the breakfast chapter for more unusual popcorn recipes.

## Quality Popping Corn

Corn pops when the kernel is heated rapidly enough that the moisture inside the kernel quickly turns to steam (which takes up more room than liquid) and literally explodes the tight seed coat of the kernel. For the greatest popping expansion, moisture content should be between 13% and 14.5%. Kernels that are too dry may pop only partly open, if at all. Kernels with too much moisture pop loudly, but the pieces are tough and jagged. A batch of popcorn with about 8-12 unpopped kernels after cooking is considered a good popping. If there are many more than this, consider a problem with either your cooking technique or your popcorn moisture content.

Since popcorn usually has the proper moisture when bought, the challenge is to keep that moisture at the same level. A sealed can, glass or plastic container should do this for years. The plastic bags that popping corn is sometimes sold in should also hold moisture well if the bag has no holes. After opening, the popcorn from these bags should be transferred to an airtight container.

## My Popcorn is Too Dry!

If your popcorn is not popping well and you have reason to believe the popcorn is too dry, you may try raising the moisture content by the following method.

Place 10 ounces (liquid measure) popcorn kernels and 1 teaspoon water in an airtight container. Shake the container three times a day for 2-3 days before using. For larger amounts, use 1 tablespoon water with one quart popcorn kernels. This should raise the moisture content by about 1.5%.

## Backyard Popcorn

Method 1: Use a commercial campfire popcorn popper and follow manufacturer's directions.

Method 2: Use 12-inch square piece of foil, 1 T. oil, 1 T. popcorn kernels, and salt to taste. Place oil and popcorn in center of foil. Bring together corners and twist to make a pouch. Securely attach a stick to the pouch with a wire or string. Holding stick, hang pouch over fire or coals. Begin shaking as soon as popcorn starts to pop. Remove just before popping stops. Open carefully to avoid steam burns. Salt to taste. Makes 1 serving.

Method 3: Use skillet or large pot with lid over fire or coals. Cook as for stovetop method in recipe section.

## Ground Popcorn

Popped popcorn may be ground to three different textures for different purposes.

COARSELY GROUND POPCORN: Use hand to break apart larger pieces of popcorn. About 1 ½ c popcorn will make 1 c coarsely ground popcorn. Useful as a topping to soups and stews instead of crackers.

MEDIUM-GROUND POPCORN: This requires the use of a blender (if you have electricity) or a hand-operated meat grinder with a coarse blade. If blender is used, do about a handful at a time and use blender on low for 20 seconds. About 1 ½ c popcorn will make 1 c medium-ground popcorn. Useful to make popcorn patties or loaf, or in bread or muffins.

FINELY GROUND POPCORN: This requires the use of a blender or a hand-operated meat grinder with a fine blade. In blender, several handfuls of popcorn can be done at a time, and grinding should continue on low speed for about 2 minutes. About 2 c of popcorn will make 1 c finely ground popcorn. Useful to top casseroles, instead of breadcrumbs as a coating or topping, to stuff peppers.

# Basic Popcorn

Makes 4 servings. Prep Time: 5 minutes

**¼ c oil**
**$^2/_3$ c unpopped popcorn**
**½ tsp salt, or to taste**

1. Using a deep skillet or 4-6 quart pot, set heat to high. Heat together oil and popcorn.
2. Put on screen or lid so steam is allowed to escape.
3. Begin to shake the pan as soon as popping starts. This keeps the popped corn from burning while the other kernels have a chance to pop.
4. Remove from heat just BEFORE popping ends, to avoid scorching. This all happens very quickly, and may take only 30-60 seconds total popping time. Pour into serving bowl. Sprinkle with salt and toss.

**Dry Pop Popcorn**: Omit oil. Use a deep, heavy pot or pressure cooker (no pressure lid) with a nonstick surface. Heat pot for 1 minute over medium heat before adding popcorn. Cover with screen or lid to let steam escape. Shake until popping almost stops. Remove from heat.

CAUTION: Allow pot (either recipe) to cool in a safe place after popping corn. DO NOT rinse with cold water while hot, as pot may warp.

*Remember that popcorn is great as a topping for soups, stews and chili, to use in place of the usual soda crackers!*

---

Per serving (excluding unknown items): 194.1 Calories; 14.4g Fat (65.1% calories from fat); 1.7g Protein; 15.6g Carbohydrate; 0mg Cholesterol; 274mg Sodium. Exchanges: 1 Grain(Starch); 2 ½ Fat.

# Popcorn: Theme and Variations

Makes 4 servings. Prep Time: 5 minutes

---

**4 qt popcorn**

**Chicken**: 1 tsp chicken bouillon granules.

**Herbed**: Salt to taste, ¾-1 ½ tsp oregano, basil or other herb.

**Mexicali**: ½ tsp garlic powder, ½ tsp chili powder, ¼ tsp cumin, ½ tsp parsley flakes, salt to taste.

**Peanut Butter**: 2-3 tbsp peanut butter, melted over low heat until smooth. Pour over popcorn and mix well.

**Simon And Garfunkel**: ½ tsp parsley flakes, ⅛ tsp sage, ¼ tsp crushed rosemary, ½ tsp thyme.

**Sprouts**: 1 c alfalfa or ¾ c bean sprouts, chopped; salt to taste.

**Mustard**: Stir together 1 tbsp prepared mustard and 1-2 tbsp oil. Toss with popcorn.

**Caliente**: Combine 3 or more drops Tabasco sauce, 2 tbsp oil, ¾ tbsp chili powder, 1 tsp cumin, ½ tsp coriander, ¼ tsp cayenne, ¼ tsp garlic powder, ½ tsp salt. Pour over popcorn and toss.

**Chicken-Little**: 1 tbsp chicken bouillon granules, 2 tsp minced onion flakes, 2 tsp marjoram, 2 tsp pasley flakes, 1 tsp sage, 1 tsp celery seed, salt to taste.

**Worcestershire**: Combine 2 tbsp oil, 2 tbsp Worcestershire sauce, ½ tsp garlic powder, 2 tbsp basil. Pour over popcorn and mix well.

**Pb And J**: Melt together 1 tbsp oil, 3 tbsp peanut butter, 2 tbsp jelly over low heat. Pour over popcorn and toss.

**Sugar And Spice**: Combine 2 tbsp oil, 1 tbsp water, 4 tbsp sugar, ½ tsp nutmeg and 1 tsp cinnamon. Pour over popcorn and toss.

# Gingersnap Popcorn Crunch

Makes 6 servings. Prep Time: 40 minutes

**12 c popcorn**
**3 tbsp shortening**
**3 tbsp oil**
**3 tbsp water**
**1 c brown sugar**
**¼ c molasses**
**½ tsp salt**
**¼ tsp cinnamon**
**¼ tsp nutmeg**
**¼ tsp ginger**

1. Place popped popcorn in large, greased baking pan in a 250 degree oven to keep warm.
2. In a large saucepan, melt shortening and add remaining ingredients. Cook over medium heat until boiling, stirring constantly.
3. Cook to 260 degrees on candy thermometer (or see below).
4. Remove popcorn from oven. Slowly pour mixture on top and stir to coat popcorn.
5. Return to oven and bake 20 minutes, stirring once. Cool and break into pieces.

NOTE: 260 degrees is still hard ball stage, but slightly firmer than 250 degrees.

Per serving (excluding unknown items): 402.1 Calories; 19.5g Fat (42.2% calories from fat); 2.0g Protein; 57.8g Carbohydrate; 0mg Cholesterol; 392mg Sodium. Exchanges: 1 Grain(Starch); 4 Fat; 3 Other Carbohydrates.

# Honey-Popcorn Balls

Makes 8 servings. Prep Time: 20 minutes

**1 $^1/_3$ c honey** **8 c popcorn**

1. Place popped popcorn in a large, greased baking pan in 250 degree oven to keep warm.
2. Heat honey (or 1 c honey and 1 c sugar) on stove to 250 degrees (use candy thermometer, or see below).
3. Remove popcorn from oven and pour honey on top. Allow to cool slightly.
4. Shape into 8 balls.

NOTE: 250 degrees is "hard ball" stage in candy-making. Pour a small amount of syrup from a spoon into a bowl of cold water. Syrup can form a ball that holds its shape, even when pressed, but is still slightly pliable.

*Traditional popcorn ball recipes all use corn syrup. If you have this on hand, you may use any recipe from a traditional cookbook.*

Per serving (excluding unknown items): 226.8 Calories; 3.1g Fat (11.4% calories from fat); 1.2g Protein; 52.8g Carbohydrate; 0mg Cholesterol; 100mg Sodium. Exchanges: ½ Grain(Starch); ½ Fat; 3 Other Carbohydrates.

# Molasses Crunch

Makes 6 servings. Prep Time: 25 minutes

**12 c popcorn**
**1 $^{1}/_{3}$ c molasses**
**1 c sugar**
**2 tbsp shortening**
**1 tbsp vinegar**
**$^{2}/_{3}$ c water**
**¼ tsp baking soda**
**1 tsp vanilla**

1. Place popcorn in large, greased baking pan in a 250 degree oven to keep warm. Grease 2 8x8 pans.
2. Combine molasses, sugar, shortening, vinegar, and water. Cook SLOWLY without stirring until mixture reaches 250 degrees on candy thermometer (hard ball stage). Add baking soda and stir well.
3. Remove from heat and stir in vanilla.
4. Remove popcorn from oven. Pour syrup on top and mix well. Turn into greased pans. Cut when cool into 24 squares.

---

Per serving (excluding unknown items): 473.3 Calories; 10.5g Fat (19.4% calories from fat); 2.0g Protein; 96.4g Carbohydrate; 0mg Cholesterol; 275mg Sodium. Exchanges: 1 Grain(Starch); 2 Fruit; 2 Fat; 5 ½ Other Carbohydrates.

# Popcorn Energy Bars

Makes 12 servings. Prep Time: 1 hour

**½ c shortening**
**½ c oil**
**1 c brown sugar**
**2 med eggs, reconstituted**
**2 tbsp water**
**1 tsp vanilla**
**1 ¾ c whole wheat flour**
**1 tsp baking soda (see note)**
**½ tsp salt**
**1 ½ c raisins**
**2 c popcorn, medium-ground**
**3 c zucchini, coarsely grated**
**1 tsp cinnamon**

1, Preheat oven to 350 degrees. Grease a 9x13 baking pan.
2. In a large mixing bowl, cream shortening, oil and brown sugar until fluffy. Beat in eggs, water and vanilla.
3. Stir together flour, baking soda and salt. Add to creamed mixture and stir just until blended.
4. Stir in raisins, ground popcorn and zucchini. Spread in pan. Sprinkle with cinnamon.
5. Bake 35-40 minutes, until wooden pick inserted in center comes out clean.
6. Cut into 24 bars when cool.

Per serving (excluding unknown items): 366.2 Calories; 19.4g Fat (45.8% calories from fat); 4.6g Protein; 47.2g Carbohydrate; 36mg Cholesterol; 232mg Sodium. Exchanges: 1 Grain(Starch); 1 Fruit; 4 Fat; 1 Other Carbohydrates.

# Popcorn Patties

Makes 6 servings. Prep Time: 45 minutes

**2 c bread crumbs**
**2 c popcorn, medium-ground**
**1 c milk, reconstituted**
**2 med eggs, reconstituted**
**salt and pepper, to taste**
**6 tbsp whole wheat flour**
**4 tbsp shortening, for frying**

1. In a large bowl, combine all ingredients except flour and shortening. Form into 6 patties (see Lentil Burger recipe for instructions). Use flour instead of bread crumbs for coating.
2. Melt shortening in a large skillet. Brown patties, turning to cook both sides, about 5 minutes per side.

*Milk and eggs provide complementary protein with the popcorn and wheat, when used in this recipe.*

Serve as hamburgers, or garnished with tomato sauce.

Per serving (excluding unknown items): 315.6 Calories; 15.1g Fat (43.0% calories from fat); 9.1g Protein; 36.1g Carbohydrate; 77mg Cholesterol; 384mg Sodium. Exchanges: 2 Grain(Starch); ½ Lean Meat; 2 ½ Fat.

# Raisin-Nut Crunch

Makes 6 servings. Prep Time: 1 hour

**8 c popcorn**
**1 c chopped nuts**
**2 tbsp shortening**
**1 tbsp oil**
**1 c sugar**
**1/3 c honey**
**½ c water**
**½ tsp salt**
**1 c raisins**

1. Keep popcorn and nuts warm in large, greased baking pan in a 250 degree oven.
2. In a large saucepan, melt shortening. Add remaining ingredients except raisins. Cook over medium heat, stirring constantly until mixture starts to boil. Cook to 250 degrees on candy thermometer (hard ball stage.)
3. Remove popcorn from oven. Stir in raisins.
4. Pour syrup slowly on top and mix. Spread 1-inch deep in pan and bake 45 minutes, stirring occasionally.
5. Cool and break apart into chunks.

Per serving (excluding unknown items): 535.9 Calories; 24.1g Fat (38.3% calories from fat); 6.1g Protein; 81.4g Carbohydrate; 0mg Cholesterol; 314mg Sodium. Exchanges: 1 Grain(Starch); ½ Lean Meat; 3 ½ Fruit; 4 ½ Fat; 3 ½ Other Carbohydrates.

# UNEXPECTED DELIGHTS

This chapter provides recipes for cookies, cakes, pies and desserts. The chapter title was selected because the "sweets" we are so accustomed to may, indeed, become less frequent and more appreciated if we will have to spend more time preparing meals from scratch and if ingredients become scarce. If no electric mixer is used, all mixing and creaming for cakes and cookies will need to be done by hand, usually using a sturdy rotary eggbeater or a large wooden spoon and lots of elbow grease.

Whole wheat flour may be substituted, cup for cup, in most cookie recipes the reader is already familiar with, such as chocolate chip or oatmeal raisin. If the results are cookies that are thin and crumbly, more whole wheat flour (or soft white flour, or all-purpose flour, if available) should be added to achieve desired consistency. Whole wheat flour, especially if coarsely ground, may not absorb as much liquid as all-purpose flour, or it may just take a longer time for it to absorb the liquid. Whole wheat may be used in more delicate cookies such as spritz or cutout cookies, but all-purpose or soft white wheat flour would be better for these recipes.

The same principles hold true for cake recipes. Angel food cakes, sponge cakes or other light-textured cakes may not result as desired if whole wheat flour is used, although they would still be edible. Average-textured cakes will be a little heavier and more crumbly than usual, but certainly acceptable.

Piecrusts, being a pastry, are also more challenging with whole wheat flour. Using all-purpose of soft white wheat flour is easier, but, with practice, a crisp, intact whole wheat crust can be made—and it is worth the effort! For those who may want to bypass the piecrust rolling process, several recipes for a graham cracker-type pressed crust are also provided.

Baking cookies may require some changes "off the grid." Since baking without electricity may put tight constraints on oven space and time, instructions for cookies will be given for making bars as well as the usual round shape. In this way, using a 9x13 pan for example, 24 cookies could be baked at once in a very small space.

# Apple Fritters

Makes 8 servings. Prep Time: 45 minutes

**4 c whole wheat flour**
**2 tsp baking soda**
**1 tsp salt**
**4 tbsp sugar**
**½ tsp nutmeg**
**4 med eggs, reconstituted**
**2 ½ c milk, reconstituted**
**4 tbsp oil**
**4 lg apples, chopped**

1. Stir together dry ingredients. Stir in oil, milk and eggs. Stir in chopped apples. Canned corn or any other fruit may be substituted for the apples.
2. Heat about 2-inches oil in deep skillet to 375 degree ( a one-inch cube of bread will brown in about 25 seconds).
3. Drop by spoonfuls into hot oil. Turn once to brown on both sides. Drain on brown paper or paper towels.

Per serving (excluding unknown items): 413.4 Calories; 13.3g Fat (27.6% calories from fat); 14.0g Protein; 64.2g Carbohydrate; 118mg Cholesterol; 653mg Sodium. Exchanges: 3 Grain(Starch); ½ Lean Meat; 1 Fruit; 2 ½ Fat; ½ Other Carbohydrates.

# Apple Pudding

Makes 12 servings. Prep Time: 1 hour

**2 c whole wheat flour**
**1 tbsp baking powder**
**1 ½ tsp cinnamon**
**1 tsp nutmeg**
**½ tsp salt**
**2 c sugar**
**¼ c soy flour**
**⅓ c shortening**
**½ c water**
**4 c apples, grated**

1. Stir together dry ingredients. Cream together sugar and shortening.
2. Stir apples into creamed mixture. Add water alternately with dry ingredients and mix well.
3. Pour into greased 9x13 pan. Bake at 350 degrees 40-45 minutes, until set and brown. Makes one 9x13 pan.

Per serving (excluding unknown items): 278.8 Calories; 6.6g Fat (20.4% calories from fat); 3.4g Protein; 54.7g Carbohydrate; 0mg Cholesterol; 182mg Sodium. Exchanges: 1 Grain(Starch); 2 ½ Fruit; 1 ½ Fat; 2 ½ Other Carbohydrates.

# Baked Apples

Makes 4 servings. Prep Time: 45 minutes

**4 lg apples**
**8 tbsp sugar**
**½ tsp cinnamon**
**4 tsp shortening**

1. Wash and core baking apples. Place apples upright in an ungreased baking dish.
2. Fill centers of apples with a mixture of 1-2 tbsp white or brown sugar, 1/8 tsp cinnamon, and 1 tsp shortening.
3. Pour water, ¼-inch deep, in baking dish.
4. Bake, uncovered, in a 375 degree oven until tender, 30-40 minutes.
5. Spoon syrup in dish over apples several times during baking if desired.
6. Serve warm or cool. Granola or raisins may be added to filling if desired.

Per serving (excluding unknown items): 216.7 Calories; 4.8g Fat (18.8% calories from fat); 0.3g Protein; 46.3g Carbohydrate; 0mg Cholesterol; 0mg Sodium. Exchanges: 3 Fruit; 1 Fat; 1 ½ Other Carbohydrates.

# Cooked Dried Fruits

Makes 1 serving. Prep Time: 45 minutes

**any dried fruit**
**water**
**sugar**

1. Place fruit in appropriate size saucepan.
2. Add enough cold water just to cover. Soak about 30 minutes, until fruit plumps.
3. Simmer in same water in a tightly covered pan until tender, 30-45 minutes.
4. Sweeten to taste with sugar.
5. Serve warm or cool.

Per serving (excluding unknown items): 0.0 Calories; 0.0g Fat (0.0% calories from fat); 0.0g Protein; 0.0g Carbohydrate; 0.0mg Cholesterol; 0.0mg Sodium. Exchanges: Free.

# Creeping Crust Cobbler

Makes 9 servings. Prep Time: 45 minutes

**1/3 c shortening**
**1 c whole wheat flour**
**1 c sugar**
**1 tsp baking powder**
**1 ½ tbsp powdered milk**
**2/3 c water**
**1 c sugar**
**2 c raspberries**

1. Melt shortening in a 9x9 pan; spread to cover.
2. Mix flour, 1 c sugar, baking powder, and dry milk. Add water and mix. Spoon over melted shortening.
3. Heat fresh of canned fruit with remaining 1 c sugar, adjusting sugar according to the fruit's sweetness.
4. Pour over dough. Bake at 350 degrees about 30 minutes, until crust is golden brown.
5. Crust will rise to the top.

*Same fruits may be used as in Fruit Crisp recipe. This fun cobbler starts with the crust on the bottom, but by the end of baking time it has crept to the top!*

Per serving (excluding unknown items): 304.6 Calories; 8.4g Fat (23.8% calories from fat); 2.4g Protein; 57.9g Carbohydrate; 1mg Cholesterol; 47mg Sodium. Exchanges: ½ Grain(Starch); 3 Fruit; 1 ½ Fat; 3 Other Carbohydrates.

## Fresh Fruit Compote

Makes 6 servings. Prep Time: 25 minutes

**8 ozs jelly**
**½ c water**
**2 lg pears, cut in half**
**2 lg peaches, cut in half**
**4 lg plums, cut in half**

1. Heat jelly (currant or other flavor) and water over low heat, stirring occasionally, until smooth.
2. Cover and heat to boiling. Reduce heat. Add fruits and simmer until tender, about 10 minutes.
3. Allow to cool somewhat before serving.

NOTE: Any combination of fruits may be used.

Per serving (excluding unknown items): 156.5 Calories; 0.8g Fat (4.1% calories from fat); 1.1g Protein; 39.6g Carbohydrate; 0mg Cholesterol; 5mg Sodium. Exchanges: 1 ½ Fruit; 1 Other Carbohydrates.

## Fried Apple Rings

Makes 6 servings. Prep Time: 30 minutes

**6 lg apples, cored**
**6 tbsp brown sugar**
**½ tsp cinnamon**
**2 tbsp shortening**

1. Slice unpeeled, cored, tart apples into rings about ½-inch thick.
2. Heat shortening in skillet. Mix brown sugar and cinnamon for coating.
3. Dip apple slices in mix and fry until tender, about 5 minutes.

*This in another pioneer recipe, easy to make and sure to please.*

Per serving (excluding unknown items): 171.4 Calories; 4.8g Fat (23.6% calories from fat); 0.3g Protein; 34.6g Carbohydrate; 0mg Cholesterol; 5mg Sodium. Exchanges: 1 ½ Fruit; 1 Fat; 1 Other Carbohydrates.

# Fruit Cobbler

Makes 9 servings. Prep Time: 45 minutes

**4 c canned peaches**
**1 ½ c whole wheat flour**
**2 tbsp sugar**
**2 tsp baking powder**
**½ tsp salt**
**3 tbsp powdered milk**
**3 tbsp shortening**
**1 c water**

1. Place sweetened fruit in greased 8x8 pan.
2. In another bowl, stir together dry ingredients. Cut in shortening using pastry blender or two knives until mixture resembles coarse crumbs.
3. Gently stir in water. Spread topping over fruit.
4. Bake at 375 degrees 25-30 minutes, until fruit is tender and topping is golden brown.
5. Makes one 8x8 pan.

*Same fruits may be used as for Fruit Crisp. Sweeten to taste, as topping itself has little sugar.*

---

Per serving (excluding unknown items): 214.2 Calories; 5.5g Fat (21.4% calories from fat); 4.0g Protein; 41.2g Carbohydrate; 3mg Cholesterol; 218mg Sodium. Exchanges: 1 Grain(Starch); 1 ½ Fruit; 1 Fat.

# Fruit Crisp

Makes 9 servings. Prep Time: 45 minutes

**4 c apple, chopped**
**½ c whole wheat flour**
**½ c rolled oats**
**$^{2}/_{3}$ c brown sugar, or white sugar**
**1 tsp cinnamon**
**¼ c shortening, or oil**

1. Place chopped apples in a greased 8x8 pan.
2. Combine remaining ingredients in a separate bowl and cut in shortening (or mix in oil) until mixture resembles coarse crumbs.
3. Sprinkle over fruit and bake at 350 degrees for 30-35 minutes.
4. Serve warm or cool. Makes one 8x8 pan.

**Granola-Fruit Crisp**: Use 1 ½ c granola instead of topping ingredients. Bake as directed.

*Most canned or fresh fruits may be used in this recipe, such as apples, apricots, blueberries, cherries, peaches, plums, raspberries, or rhubarb. Rhubarb may require twice as much sugar for sweetener.*

Per serving (excluding unknown items): 181.1 Calories; 6.3g Fat (29.9% calories from fat); 1.7g Protein; 31.4g Carbohydrate; 0mg Cholesterol; 7mg Sodium. Exchanges: ½ Grain(Starch); ½ Fruit; 1 Fat; 1 Other Carbohydrates.

# Raspberry Dumplings

Makes 6 servings. Prep Time: 50 minutes

**4 c fresh raspberries**
**1 c water**
**1 c sugar**
**1 tbsp oil**
**1 recipe Dumplings, uncooked**

1. Combine berries, water, sugar and oil. Boil over low heat until berries are soft.
2. Meanwhile, prepare sweet dumpling batter. Place on top of cooking berries in 6 portions.
3. Cover tightly and simmer 30 minutes. Serve each a dumpling and portion of berries.

*Any other berry or soft fruit may be substituted in this pioneer recipe. Adjust sugar to fruit's tartness.*

Per serving (excluding unknown items): 189.3 Calories; 2.7g Fat (12.3% calories from fat); 0.8g Protein; 42.8g Carbohydrate; 0mg Cholesterol; 2mg Sodium. Exchanges: 2 ½ Fruit; ½ Fat; 2 Other Carbohydrates.

# Stewed Apples

Makes 6 servings. Prep Time: 25 minutes

**4 lg apples, chopped**
**½ c sugar**
**½ c water**

1. Heat water and sugar to boiling, stirring occasionally.
2. Reduce heat. Add fruit. Simmer uncovered until tender, about 10 minutes. Adjust sugar according to fruit's tartness.

*Firm fruits, such as apples, pears or rhubarb may be used.*

Per serving (excluding unknown items): 118.8 Calories; 0.3g Fat (2.4% calories from fat); 0.2g Protein; 30.7g Carbohydrate; 0mg Cholesterol; 1mg Sodium. Exchanges: 2 Fruit; 1 Other Carbohydrates.

# Stewed Fruit

Makes 8 servings. Prep Time: 30 minutes

**2 lb plums**
**¾ c sugar**
**1/8 tsp salt**
**dash allspice**
**2 sticks cinnamon**
**2 c water**

1. Heat all but fruit in a 3-quart saucepan. Bring to a boil.
2. Add fruit that has been washed, halved and pitted.
3. Cook uncovered over medium heat just until tender, 10-15 minutes.
4. Cool before serving.

*Soft fruits, such as plums, peaches or apricots may be used.*

Serve as a breakfast fruit or dessert.

Per serving (excluding unknown items): 135.7 Calories; 0.7g Fat (4.5% calories from fat); 0.9g Protein; 34.0g Carbohydrate; 0mg Cholesterol; 36mg Sodium. Exchanges: 2 Fruit; 1 ½ Other Carbohydrates.

# Instant Pudding

Makes 4 servings. Prep Time: 10 minutes

**1/3 c Instant ClearJel**
**½ c sugar**
**1/8 tsp salt**
**2 c milk, reconstituted**
**2 tsp vanilla**

1. Whisk together sugar and Instant ClearJel and salt in a separate bowl.
2. Stir together milk and vanilla in a 1-quart bowl.
3. Slowly sprinkle in sugar mixture while whisking constantly to prevent lumping.
4. Pudding will set up almost immediately. Spoon into serving dishes.

*Since this contains milk, it is susceptible to spoilage and should be made just before serving.*

Per serving (excluding unknown items): 178.7 Calories; 4.1g Fat (20.6% calories from fat); 4.0g Protein; 31.4g Carbohydrate; 17mg Cholesterol; 127mg Sodium. Exchanges: ½ Non-Fat Milk; 1 ½ Fruit; ½ Fat; 1 ½ Other Carbohydrates.

# Poor Man's Pudding

Makes 8 servings. Prep Time: 1 hour and 10 minutes

**$^{1}/_{3}$ c brown sugar**
**1 c whole wheat flour**
**1 tsp baking powder**
**1 pinch salt**
**½ c milk, reconstituted**

**1 c brown sugar**
**2 c hot water**
**1 tbsp oil**
**½ tsp nutmeg**

1. Combine $^{1}/_{3}$ c. brown sugar, flour, baking powder, salt and milk to form thick batter. Spread in a greased 9x9 or round layer cake pan.
2. Mix together remaining ingredients and pour over batter. Bake at 350 degrees for about 30 minutes. Batter will rise to the top to form crust and liquid becomes pudding-like at the bottom. Spoon out some of both when serving.

**Raisin Pudding**: Add 1 cup raisins to the batter before spreading in pan.

*This recipe from the Depression days takes something very simple and turns it into something very special!*

Per serving (excluding unknown items): 214.2 Calories; 2.5g Fat (10.2% calories from fat); 2.6g Protein; 47.5g Carbohydrate; 2mg Cholesterol; 86mg Sodium. Exchanges: ½ Grain(Starch); ½ Fat; 2 ½ Other Carbohydrates.

# Vanilla Pudding

Makes 4 servings. Prep Time: 25 minutes

**$^{1}/_{3}$ c sugar**
**2 tbsp cornstarch**
**$^{1}/_{8}$ tsp salt**
**2 c milk, reconstituted**
**1 ½ tbsp soy flour**
**1 tbsp shortening**
**2 tsp vanilla**

1. Mix together sugar, cornstarch, salt and soy flour. Stir in milk gradually with a whisk.
2. Cook over medium heat (or a double boiler), stirring constantly until mixture thickens and boils.
3. Boil and stir 2-3 minutes. Remove from heat.
4. Stir in shortening (or may be omitted) and vanilla.
5. Pour into individual dishes and cool before serving.

**Butterscotch Pudding**: Substitute $^{2}/_{3}$ c brown sugar for white sugar. Decrease vanilla to 1 tsp

**Chocolate Pudding**: Increase sugar to ½ c and stir in $^{1}/_{3}$ c cocoa into sugar mixture in step #1.

*Since this contains milk, it is susceptible to spoilage and should be made shortly before serving.*

Per serving (excluding unknown items): 198.6 Calories; 7.7g Fat (35.0% calories from fat); 4.7g Protein; 27.4g Carbohydrate; 17mg Cholesterol; 127mg Sodium. Exchanges: ½ Grain(Starch); ½ Non-Fat Milk; 1 Fruit; 1 ½ Fat; 1 Other Carbohydrates.

# Apple Dumplings

Makes 6 servings. Prep Time: 1 hour

**2 ea pie crust (9 inch)**
**6 med apples**
**3 tbsp raisins**
**3 tbsp chopped nuts, optional**
**2 c brown sugar**
**1 c water**

1. Prepare pie crusts as one batch. Roll into a rectangle, 14 x 21 inches. Cut into 6 7-inch squares.
2. Core apples. Place one on each square. Mix raisins and nuts. Use extra raisins if omitting nuts. Fill each apple.
3. Moisten corners of pastry squares. Bring opposite corners up over apple and pinch. Repeat with remaining corners, and pinching edges together to seal.
4. Place in ungreased 7x12 or 9x9 baking dish. Heat brown sugar and water to boiling. Pour around dumplings.
5. Bake, spooning syrup over dumplings several times during baking, until crust is golden and apples are tender, about 40 minutes. Serve with syrup in bowls.

Per serving (excluding unknown items): 665.3 Calories; 19.2g Fat (25.2% calories from fat); 4.7g Protein; 123.8g Carbohydrate; 0mg Cholesterol; 420mg Sodium. Exchanges: 2 Grain(Starch); 1 ½ Fruit; 3 ½ Fat; 5 Other Carbohydrates.

# Applesauce Cookies

Makes 42 servings. Prep Time: 45 minutes

**1 c shortening**
**2 c brown sugar**
**3 tbsp soy flour**
**¼ c water**
**2 c applesauce**
**3 ½ c whole wheat flour**
**1 tsp baking soda**
**1 tsp nutmeg**
**1 tsp cinnamon**
**1 tea cloves**
**1 c raisins**
**½ c chopped nuts**

1. Cream shortening and sugar. Add soy flour, water and applesauce. Mix well.
2. Stir in dry ingredients. Cool as much as possible. If dough is still too soft, add a small amount of extra flour, being careful not to get the dough too dry.
3. Heat oven to 400 degrees. Drop by rounded spoonfuls onto lightly greased cookie sheet or fill two greased pans, 8x8 and 9x13. Bake 9-12 minutes for cookies or 25-30 minutes for bars.
4. Makes 7 dozen.

*Any pureed fruit may be used.*

Per serving (excluding unknown items): 149.0 Calories; 6.2g Fat (35.6% calories from fat); 1.9g Protein; 23.3g Carbohydrate; 0mg Cholesterol; 36mg Sodium. Exchanges: ½ Grain(Starch); ½ Fruit; 1 Fat; ½ Other Carbohydrates.

# Blonde Brownies

Makes 48 servings. Prep Time: 45 minutes

**2 c whole wheat flour**
**¼ tsp baking soda**
**1 tsp baking powder**
**$^{2}/_{3}$ c shortening**
**3 tbsp soy flour**
**½ c water**
**2 c brown sugar**
**2 tsp vanilla**
**1 c chocolate chips**
**$^{1}/_{3}$ c chopped nuts, optional**

1. Preheat oven to 350 degrees.
2. Melt shortening in a 3-quart saucepan. Remove from heat. Stir together soy flour and water in separate bowl.
3. Add sugar, soy flour mixture and vanilla to melted shortening.
4. Add flour, soda and baking powder gradually, mixing well.
5. Sprinkle with chocolate chips and nuts. Bake at 350 degrees for30 minutes.
6. Cool before cutting into 48 bars. Makes one 9x13 pan.

Per serving (excluding unknown items): 107.0 Calories; 5.0g Fat (39.5% calories from fat); 1.2g Protein; 15.9g Carbohydrate; 0mg Cholesterol; 19mg Sodium. Exchanges: ½ Grain(Starch); 1 Fat; 1 Other Carbohydrates.

# Brownies

Makes 16 servings. Prep Time: 45 minutes

**½ c shortening**
**1 c sugar**
**6 tbsp cocoa**
**3 tbsp soy flour**
**½ tsp baking powder (see note)**
**¼ c water**
**½ tsp vanilla**
**¾ c whole wheat flour**
**½ tsp salt**
**½ c chopped nuts, optional**

1. Melt shortening in 2-quart pan over low heat. Remove from heat.
2. Stir in sugar, cocoa, water and vanilla.
3. Stir in remaining ingredients and spread in a greased 8x8 pan.
4. Bake 30-35 minutes at 350 degrees. Cool slightly before cutting into 16 bars.
5. Makes one 8x8 pan.

NOTE: Baking powder produces cake-like brownies. For denser, fudge-type brownies, omit baking powder.

---

Per serving (excluding unknown items): 160.9 Calories; 9.5g Fat (50.1% calories from fat); 2.3g Protein; 19.0g Carbohydrate; 0mg Cholesterol; 80mg Sodium. Exchanges: ½ Grain(Starch); 1 Fruit; 2 Fat; 1 Other Carbohydrates.

# Garden Cookies

Makes 30 servings. Prep Time: 45 minutes

**1 c shortening**
**¾ c brown sugar**
**¾ c sugar**
**3 tbsp soy flour**
**½ c water**

**1 ½ c whole wheat flour**
**½ tsp salt**
**2 tsp baking powder**
**2 c oatmeal**
**1 c grated carrots**

1. Cream shortening and sugars. Add soy flour, water and vanilla. Mix well.
2. Blend in flour, salt and baking powder. Stir in oatmeal and carrots. Let dough sit 10-15 minutes.
3. Drop by spoonfuls onto greased cookie sheets or spread in greased pan, 8x8 and 8x11.
4. Bake at 375 degrees, 10 minutes for cookies or 25 minutes for bars. Makes 5 dozen cookies or bars.

*These cookies use vegetables from the garden! Try substituting grated zucchini or grated apple for the carrot.*

Per serving (excluding unknown items): 145.6 Calories; 7.4g Fat (44.4% calories from fat); 1.9g Protein; 18.9g Carbohydrate; 0mg Cholesterol; 118mg Sodium. Exchanges: ½ Grain(Starch); ½ Fruit; 1 ½ Fat; ½ Other Carbohydrates.

# Gingerbread Cookies

Makes 30 servings. Prep Time: 1 hour

---

**1 ½ c molasses**
**1 c brown sugar**
**$^{2}/_{3}$ c cold water**
**$^{1}/_{3}$ c shortening**
**7 c all-purpose flour (see note)**
**2 tsp baking soda**
**1 tsp salt**
**1 tsp allspice**
**1 tsp cinnamon**
**2 tsp ginger**
**1 tsp cloves**

1. Mix molasses, brown sugar, water and shortening.
2. Mix in remaining ingredients thoroughly. Chill if possible.
3. Roll dough ¼-inch thick on floured surface. Cut with floured cookie cutters. Place about 2 inches apart on a lightly greased cookie sheet.
4. Bake in a 350 degree oven until set, about 10-12 minutes. If preferred, dough may be rolled ½-inch thick and cookies should be bake about 15 minutes. Cookies may be frosted and/or decorated.
5. Makes about 30 large cookies.

NOTE: Soft white wheat flour or half all-purpose and half whole wheat flour may be substituted for the all-purpose flour.

*Even with no electricity, if these traditional cookies are made at Christmas time, maybe the weather will be right for chilling the dough before rolling.*

---

Per serving (excluding unknown items): 198.5 Calories; 2.6g Fat (11.8% calories from fat); 3.0g Protein; 40.9g Carbohydrate; 0mg Cholesterol; 165mg Sodium. Exchanges: 1 ½ Grain(Starch); ½ Fat; 1 Other Carbohydrates.

# Kitchen Sink Cookies

Makes 36 servings. Prep Time: 45 minutes

**1 c shortening**
**1 c sugar**
**1 c brown sugar**
**½ c milk, reconstituted**
**1 tbsp vanilla**
**3 tbsp soy flour**
**2 ½ c whole wheat flour**
**1 tbsp baking powder**
**2 ½ c rolled oats**
**1 c raisins**
**1 c chopped nuts**
**1 c chocolate chips**

1. Cream shortening and sugars. Add milk, soy flour and vanilla.
2. Add dry ingredients and blend thoroughly.
3. Drop by spoonfuls onto ungreased cookie sheet, or spread in two greased pans, 9x9 and 8x11 or equivalent.
4. Bake in a 350 degree oven, 10-12 minutes for cookies or 30 minutes for bars.
5. Cool slightly before cutting bars. Makes 6 dozen cookies or bars.

*These cookies have everything in them but the kitchen sink!*

---

Per serving (excluding unknown items): 216.1 Calories; 10.5g Fat (41.6% calories from fat); 3.4g Protein; 29.8g Carbohydrate; 0mg Cholesterol; 37mg Sodium. Exchanges: ½ Grain(Starch); ½ Fruit; 2 Fat; 1 Other Carbohydrates.

# Molasses Crinkles

Makes 24 servings. Prep Time: 45 minutes

**1 c brown sugar**
**¾ c shortening**
**¼ c molasses**
**¼ c water**
**1 ½ tsp soy flour**
**2 ¼ c whole wheat flour**
**2 tsp baking soda**
**1 tsp cinnamon**
**1 tsp ginger**
**½ tsp cloves**
**¼ c sugar, for topping**

1. Cream brown sugar, shortening and molasses. Mix in water and soy flour.
2. Stir in dry ingredients and cool as much as possible. Heat oven to 375 degrees.
3. Make one cookie into 1 ¼-inch ball. Bake test cookie. If it spreads too much, add a small amount of flour to remaining dough. If it stays round and does not flatten and develop crinkle lines, dough is too dry and must be thinned with a small amount of water or milk.
4. To bake, shape dough into 1 ¼-inch balls. Roll in ¼ c sugar. Place balls, about 3 inches apart on lightly greased baking sheet.
5. Bake just until set, 10-12 minutes. Remove from sheet immediately. Makes about 4 dozen.

Per serving (excluding unknown items): 147.5 Calories; 6.7g Fat (39.1% calories from fat); 1.6g Protein; 21.7g Carbohydrate; 0mg Cholesterol; 111mg Sodium. Exchanges: ½ Grain(Starch); 1 ½ Fat; 1 Other Carbohydrates.

## No-Bake Chewy Goober Bars

Makes 40 servings. Prep Time: 25 minutes

**3 tbsp shortening**
**1/3 c peanut butter**
**1/3 c brown sugar**
**¾ c honey**
**¾ tsp vanilla**
**3 c rolled oats, coarsely ground**
**¾ c raisins**
**½ tsp cinnamon**

1. Grind rolled oats by using electric blender or hand meat grinder. Texture should be coarser than flour, more like commercial oat bran cereal.
2. Place shortening, peanut butter, brown sugar, honey and vanilla in a heavy saucepan.
3. Cook over low heat, stirring constantly until mixture comes to a boil. Remove from heat.
4. Combine dry ingredients in a large bowl. Pour syrup mixture over top. Stir until well-coated.
5. Press mixture firmly and evenly into a greased 9x9 pan. Let cool
6. Makes one 9x9 pan. Cut into bars 40 bars by cutting 5 x 8.

Per serving (excluding unknown items): 79.2 Calories; 2.4g Fat (26.2% calories from fat); 1.6g Protein; 13.7g Carbohydrate; 0mg Cholesterol; 12mg Sodium. Exchanges: ½ Grain(Starch); ½ Fat; ½ Other Carbohydrates.

## No-Bake Chocolate Drops

Makes 15 servings. Prep Time: 15 minutes

**1 c cocoa**
**½ c shortening**
**¼ c peanut butter**
**1 ½ c sugar**
**¼ c water**
**1tsp vanilla**
**3 c rolled oats**

1. Combine all ingredients except rolled oats in a medium saucepan and stir over medium heat until peanut butter is melted and sugar is dissolved.
2. Remove from heat. Stir in oats and mix well.
3. Form about 30 small balls and place on serving dish. Allow to cool.

Per serving (excluding unknown items): 239.4 Calories;10.8g Fat (38.0% calories from fat);4.8g Protein; 34.9g Carbohydrate; 0mg Cholesterol; 23mg Sodium. Exchanges: 1 Grain(Starch); 1 ½ Fruit; 2 Fat; 1 ½ Other Carbohydrates.

## No-Bake Chocolate Peanut Butter Bars

Makes 48 servings. Prep Time: 30 minutes

**1 ½ c peanut butter**
**1 ⅛ c brown sugar**
**1 ¾ c powdered sugar**
**4 tbsp shortening, melted**
**1 ½ c chocolate chips, melted**
**3 tbsp shortening, melted**
**1 tbsp water**

1. Beat together peanut butter, brown sugar, powdered sugar and 4 tbsp melted shortening.
2. Spread evenly in a 9x13 pan.
3. For top layer, melt together chocolate chips and 3 tbsp shortening over double boiler or low heat. Stir in 1 tbsp water.
4. Spread on top of peanut butter layer.
5. Cut into 48 bars when cool. These bars may get soft if kept too warm.

Per serving (excluding unknown items): 133.7 Calories; 8.0g Fat (50.3% calories from fat); 2.3g Protein; 15.5g Carbohydrate; 0mg Cholesterol; 41mg Sodium. Exchanges: 1 ½ Fat; 1 Other Carbohydrates.

## No-Bake Peanut Butter Cookies

Makes 24 servings. Prep Time: 20 minutes

**¾ c powdered milk**
**1 c rolled oats**
**½ c peanut butter**
**½ c honey**

1. Stir together dry milk and rolled oats.
2. Stir in peanut butter and enough honey to form small balls using hands.
3. Add raisins or crushed nuts, if desired.
4. Form into 24 balls.

Per serving (excluding unknown items): 85.9 Calories; 4.0g Fat (39.5% calories from fat); 2.9g Protein; 10.7g Carbohydrate; 4mg Cholesterol; 41mg Sodium. Exchanges: ½ Fat; ½ Other Carbohydrates.

# Oatmeal Raisin Cookies

Makes 18 servings. Prep Time: 45 minutes

**1 ½ c flour**
**1 tsp baking soda**
**1 tsp cinnamon**
**¼ tsp nutmeg**
**¼ tsp cloves**
**½ c shortening**
**1 ½ c brown sugar**
**3 tbsp soy flour**
**½ c water**
**1 tsp vanilla**
**1 ¾ c rolled oats**
**1 c raisins**

1. Cream shortening and sugar. Beat in soy flour, water and vanilla.
2. Stir in dry ingredients until well-blended.
3. Drop by spoonfuls onto lightly greased cookie sheet or spread into greased 9x13 pan.
4. Bake in a 375 degree oven, 8-9 minutes for cookies or 25-30 minutes for bars.
5. Cool slightly before cutting bars. Makes 3 dozen cookies or bars.

Per serving (excluding unknown items): 216.8 Calories; 6.5g Fat (26.4% calories from fat); 2.9g Protein; 38.0g Carbohydrate; 0mg Cholesterol; 79mg Sodium.
Exchanges: 1 Grain(Starch); ½ Fruit; 1 ½ Fat; 1 Other Carbohydrates.

## Peanut Butter Apple Bars

Makes 16 servings. Prep Time: 45 minutes

**1 ½ tbsp soy flour**
**½ c milk, reconstituted**
**¾ c brown sugar**
**1 tsp vanilla**
**½ c peanut butter**
**¾ c whole wheat flour**
**1 ½ tsp baking powder**
**½ tsp cinnamon**
**2 med apples, chopped**

1. Combine soy flour, milk, brown sugar, vanilla. Blend in peanut butter.
2. Blend in dry ingredients and apples. If dough appears too soft, add a small amount of extra flour.
3. Pour into a greased 8x8 pan. Bake at 350 degrees for 25-30 minutes.
4. Cut into 16 bars. Makes one 8x8 pan.

Per serving (excluding unknown items): 123.6 Calories; 4.5g Fat (31.3% calories from fat); 3.2g Protein; 19.2g Carbohydrate; 1mg Cholesterol; 81mg Sodium. Exchanges: ½ Grain(Starch); ½ Lean Meat; ½ Fat; ½ Other Carbohydrates.

## Peanut Butter Cookies

Makes 18 servings. Prep Time: 45 minutes

**½ c sugar**
**½ c brown sugar**
**½ c shortening**
**½ c peanut butter**
**1 ½ tbsp soy flour**
**¼ c water**
**1 ¼ c whole wheat flour**
**¾ tsp baking soda**
**½ tsp baking powder**

1. Mix sugars, shortening, peanut butter, soy flour and water thoroughly.
2. Stir in dry ingredients and cool as much as possible. If dough is still too soft to be formed into balls, a small amount of flour may be added. Add as little as possible, or cookies will be dry and tough.
3. Heat oven to 375 degrees. Shape dough into 1 ¼-inch balls. Place on ungreased cookie sheet. Flatten in crisscross pattern with fork dipped in flour. Bake 10-12 minutes. Makes 3 dozen.

Per serving (excluding unknown items): 167.2 Calories; 9.5g Fat (49.0% calories from fat); 3.1g Protein; 19.2g Carbohydrate; 0mg Cholesterol; 100mg Sodium. Exchanges: ½ Grain(Starch); ½ Fruit; 2 Fat; 1 Other Carbohydrates.

# Pumpkin Cookies

Makes 18 servings. Prep Time: 45 minutes

**1 c sugar**
**1 c mashed pumpkin**
**½ c shortening**
**2 c whole wheat flour**
**1 tsp baking powder**
**1 tsp baking soda**
**1 tsp cinnamon**
**½ c raisins**
**½ c chopped nuts, optional**

1. Cream shortening, sugar and pumpkin. Stir in dry ingredients.
2. Drop by spoonfuls onto ungreased cookies sheet or spread in two greased 8x8 pans.
3. Bake at 375 degrees, 8-10 minutes for cookies or 25-30 minutes for bars.
4. Cool slightly before cutting. Makes 36 bars.

*Any mashed winter squash may be used.*

Per serving (excluding unknown items): 180.1 Calories; 8.2g Fat (39.1% calories from fat); 2.8g Protein; 26.1g Carbohydrate; 0mg Cholesterol; 92mg Sodium. Exchanges: ½ Grain(Starch); 1 Fruit; 1 ½ Fat; 1 Other Carbohydrates.

# World-Famous Chocolate Chip Cookies

Makes 36 servings. Prep Time: 45 minutes

**¾ c shortening**
**1 c sugar**
**1 c brown sugar**
**3 tbsp soy flour**
**½ c water**
**1 tsp vanilla**
**2 c whole wheat flour**
**½ tsp salt**
**1 tsp baking powder**
**1 tsp baking soda**
**2 ½ c rolled oats, finely ground**
**2 c chocolate chips**
**1 ½ c nuts**
**4 ozs chocolate bar, grated**

1. Grind oats to powder using blender or grain grinder.
2. Cream shortening and sugars. Add soy flour, water and vanilla.
3. Stir in dry ingredients until well-blended.
4. Drop by large spoonfuls onto ungreased cookie sheets or spread in two greased pans, 9x9 and 9x13, or equivalent. Bake at 375 degrees for 10-12 minutes for cookies or 25-30 minutes for bars. Cool slightly before cutting bars.
5. Makes 36 large cookies or bars.

*See if you can guess whose famous cookies these are!*

Per serving (excluding unknown items): 240.6 Calories; 12.9g Fat (45.4% calories from fat); 3.6g Protein; 31.4g Carbohydrate; 0mg Cholesterol; 80mg Sodium. Exchanges: ½ Grain(Starch); ½ Fruit; 2 ½ Fat; 1 ½ Other Carbohydrates.

# Your Basic Chocolate Chip Cookie

Makes 30 servings. Prep Time: 45 minutes

**2 ½ c whole wheat flour**
**1 tsp baking soda**
**1 tsp salt**
**¾ c shortening**
**¾ c sugar**
**¾ c brown sugar**
**1 tsp vanilla**
**3 tbsp soy flour**
**2 c chocolate chips**
**1 c chopped nuts, optional**

1. Cream together shortening and sugars. Add vanilla, soy flour and water. Blend thoroughly.
2. Add dry ingredients and mix well.
3. Drop by spoonfuls onto ungreased baking sheets; or spread in two greased pans, 8x8 and 8x11.
4. Bake at 375 degrees, 9-11 minutes for cookies or 20-25 minutes for bars.
5. Cool slightly before removing from cookie sheets or cutting bars.
6. Makes about 5 dozen cookies.

**Double Chocolate Chip Cookies**: Reduce flour to 2 c and add ¾ c cocoa with dry ingredients.

Per serving (excluding unknown items): 222.4 Calories; 12.5g Fat (47.4% calories from fat); 3.0g Protein; 28.3g Carbohydrate; 0mg Cholesterol; 118mg Sodium. Exchanges: ½ Grain(Starch); ½ Fruit; 2 ½ Fat; 1 ½ Other Carbohydrates.

# Chocolate Tidbit Cake

Makes 8 servings. Prep Time: 45 minutes

**1 ½ c whole wheat flour**
**1 c sugar**
**¼ c cocoa**
**1 tsp baking soda**
**1 tsp vanilla**
**1 tbsp vinegar**
**⅓ c oil**
**1 c cold water**
**½ c chocolate chips**

1. Using a fork, combine dry ingredients in a greased 8x8 or 9x9 square pan or an 8-9-inch layer cake pan.
2. Make a well in the middle. Pour in liquids and stir with a fork or whisk until smooth.
3. Sprinkle with chocolate chips.
4. Bake at 350 degrees 30 minutes, until pick inserted in center comes out clean.
5. Cool on rack. Remove from pan after 10 minutes. Sprinkle with powdered sugar, if desired.
6. Makes one square or round layer cake.

*Mix-in-the-pan cakes minimize cleanup and effort!*

Per serving (excluding unknown items): 328.3 Calories; 14.0g Fat (36.0% calories from fat); 4.2g Protein; 51.9g Carbohydrate; 0mg Cholesterol; 162mg Sodium. Exchanges: 1 Grain(Starch); 1 ½ Fruit; 3 Fat; 2 ½ Other Carbohydrates.

## Depression Cake

Makes 8 servings. Prep Time: 50 minutes

**1 c raisins**
**2 c water**
**2 tbsp oil**
**1 c brown sugar**
**2 c whole wheat flour**
**1 tbsp baking powder**
**1 tsp nutmeg**
**1 tsp cinnamon**

1. Cook raisins in water and simmer until only 1 c liquid remains. Allow to cool.
2. Cream sugar and oil. Add raisins and liquid, then dry ingredients.
3. Bake in a greased 9x9 pan for 30-40 minutes at 350 degrees.

*This cake gets its name from The Great Depression, when eggs, butter and milk were scarce. It is a variation of a recipe originally from Eastern Europe, brought to America by immigrants.*

---

Per serving (excluding unknown items): 292.7 Calories; 4.2g Fat (12.1% calories from fat); 4.7g Protein; 63.7g Carbohydrate; 0mg Cholesterol; 152mg Sodium. Exchanges: 1 ½ Grain(Starch); 1 Fruit; 1 Fat; 2 Other Carbohydrates.

## Gingerbread Cake

Makes 9 servings. Prep Time: 1 hour

**2 ½ c whole wheat flour**
**1/3 c sugar**
**1 c molasses**
**1 c hot water**
**½ c shortening**
**1 ½ tbsp soy flour**
**1 tsp baking soda**
**1 tsp ginger**
**1 tsp cinnamon**
**½ tsp salt**

1. Grease and flour a 9x9 pan. Cream shortening, sugar and molasses.
2. Add dry ingredients alternately with water to creamed mixture.
3. Pour into prepared pan. Bake at 325 degrees for about 50 minutes, until toothpick inserted in center comes out clean.

Top with applesauce, if desired.

---

Per serving (excluding unknown items): 344.5 Calories; 12.3g Fat (30.7% calories from fat); 4.9g Protein; 57.3g Carbohydrate; 0mg Cholesterol; 275mg Sodium. Exchanges: 1 ½ Grain(Starch); ½ Fruit; 2 ½ Fat; 2 Other Carbohydrates.

# Honey-Applesauce Cake

Makes 12 servings. Prep Time: 1 hour

**½ c shortening**
**1 c honey**
**3 tbsp soy flour**
**2 ¼ c whole wheat flour**
**⅓ c powdered milk**
**1 tsp baking powder**
**½ tsp salt**
**½ tsp cinnamon**
**¼ tsp cloves**
**1 ½ c applesauce**
**1 c raisins**
**1 c chopped nuts, optional**
**1 c powdered sugar**
**¼ tsp vanilla**
**1 tbsp milk, reconstituted**

1. Cream shortening and honey until light and fluffy.
2. Stir together soy flour, wheat flour powdered milk, baking powder, salt and spices.
3. Add to creamed mixture alternately with applesauce. Fold in raisins and nuts.
4. Pour into a greased and floured 12-c Bundt pan. Bake at 325 degrees for 35-45 minutes.
5. Immediately turn out onto wire rack. Cool.
6. Make glaze by stirring together powdered sugar, vanilla and 1-2 tbsp milk. Drizzle over cooled cake.
7. Makes one 12-c tube (Bundt) cake.

Per serving (excluding unknown items): 435.1 Calories; 17.0g Fat (33.2% calories from fat); 7.0g Protein; 70.1g Carbohydrate; 4mg Cholesterol; 139mg Sodium. Exchanges: 1 ½ Grain(Starch); ½ Lean Meat; 1 Fruit; 3 Fat; 2 Other Carbohydrates.

# Hot Fudge Sundae Cake

Makes 8 servings. Prep Time: 50 minutes

**1 c whole wheat flour**
**2 tsp baking powder**
**¾ c sugar**
**2 tbsp cocoa**
**½ c milk, reconstituted**
**2 tbsp oil**
**¼ c chopped nuts, optional**
**1 c brown sugar**
**¼ c cocoa**
**1 ½ c hot water**

1. Combine flour, baking powder, sugar and cocoa with a fork in a greased 8x8 pan or round layer cake pan.
2. Make a well in center. Pour in milk and oil. Stir in nuts until well-blended.
3. In a small bowl, combine brown sugar and cocoa. Sprinkle evenly over batter.
4. Carefully pour hot water over all. DO NOT STIR!
5. Bake at 350 degrees 30-45 minutes, until edges are firm. Use baking pan as serving dish. Do not remove cake.
6. Serve warm. Makes one square or round layer cake.

*This mix-in-the pan cake is easy on cleanup. When done, it looks like it has chunks of chocolate cake floating in thick, rich pudding.*

Per serving (excluding unknown items): 303.5 Calories; 7.3g Fat (20.2% calories from fat); 4.1g Protein; 60.5g Carbohydrate; 2mg Cholesterol; 113mg Sodium. Exchanges: 1 Grain(Starch); 1 Fruit; 1 ½ Fat; 3 Other Carbohydrates.

# Pinto Party Cake

Makes 12 servings. Prep Time: 1 hour

**1 c sugar**
**¼ c shortening**
**1 ½ tbsp soy flour**
**2 c cooked pinto beans, mashed**
**1 c whole wheat flour**
**1 tsp baking soda**
**4 tbsp water**
**1 tsp cinnamon**
**½ tsp cloves**
**½ tsp allspice**
**3 med apples, diced**
**¼ c chopped nuts, optional**
**2 tsp vanilla**

1. Cream shortening and sugar. Add mashed beans and mix well.
2. Stir together dry ingredients. Add alternately with water and vanilla to creamed mixture.
3. Stir in apples and nuts. Pour into a lightly greased 10-inch tube or Bundt pan.
4. Bake 45 minutes at 350 degrees. Cool slightly before removing from pan.
5. Makes one Bundt cake.

*This cake may be glazed like the Honey-Applesauce Cake.*

Per serving (excluding unknown items): 222.9 Calories; 6.6g Fat (25.5% calories from fat); 4.7g Protein; 38.4g Carbohydrate; 0mg Cholesterol; 107mg Sodium. Exchanges: 1 Grain(Starch); 1 ½ Fruit; 1 Fat; 1 Other Carbohydrates.

# Plain Vanilla Cake

Makes 8 servings. Prep Time: 45 minutes

**1 ¼ c flour**
**1 c sugar**
**1 ½ tsp baking powder**
**½ tsp salt**
**3 tbsp powdered milk**
**1 c water**
**1/3 c shortening**
**1 ½ tbsp soy flour**
**1 tsp vanilla**

1. Grease and flour a round layer pan or an 8x8 or 9x9 pan.
2. Cream shortening and sugar. Stir together dry ingredients.
3. Add dry ingredients alternately to creamed mixture with water and vanilla.
4. Pour into prepared pan and bake at 350 degrees 35-40 minutes, until toothpick inserted in center comes out clean.
5. Makes one layer or square cake. Frost as desired, or use batter in Stovetop Upside-Down Cake.

*All-purpose flour or soft white wheat flour will produce a cake lighter in texture and color.*

---

Per serving (excluding unknown items): 264.7 Calories; 9.7g Fat (32.8% calories from fat); 3.2g Protein; 41.8g Carbohydrate; 3mg Cholesterol; 214mg Sodium. Exchanges: 1 Grain(Starch); 1 ½ Fruit; 2 Fat; 1 ½ Other Carbohydrates.

## Soya Applesauce Cake

Makes 12 servings. Prep Time: 1 hour

**1 ¼ c whole wheat flour**
**1 c soy flour**
**½ c powdered milk**
**4 tsp baking powder**
**2 tsp cinnamon**
**½ tsp nutmeg**
**¼ tsp cloves**
**1 c brown sugar**
**½ c oil**
**1 ¼ c water**
**1 c applesauce**

1. Cream sugar, oil, soy flour and water.
2. Stir together dry ingredients in a separate bowl.
3. Add to creamed mixture alternately with applesauce.
4. Turn into a greased and flours 8x12 pan and bake at 350 degrees for 50 minutes, until pick inserted in center comes out clean.

Per serving (excluding unknown items): 267.1 Calories; 12.3g Fat (39.7% calories from fat); 5.6g Protein; 36.4g Carbohydrate; 5mg Cholesterol; 151mg Sodium. Exchanges: 1 Grain(Starch); ½ Lean Meat; ½ Fruit; 2 ½ Fat; 1 Other Carbohydrates.

## Cocoa-Lentil Cake

Makes 12 servings Prep Time: 45 minutes

**1 ¼ c lentils, cooked, pureed**
**1 ½ c sugar**
**1 c oil**
**4 lg eggs, reconstituted**
**1tsp vanilla**
**2 c whole wheat flour**
**4 tbsp cocoa**
**1 ½ tsp baking soda**
**½ tsp salt**

1. Follow instructions for making lentil puree; measure out 1 ¼ cups. Grease and flour a 9x13 or Bundt pan.
2. Beat sugar, oil and eggs for 2 minutes. Mix in lentils and vanilla.
3. Combine dry ingredients. Add to batter; beat 2 minutes. Pour into pan.
4. Bake at 350 degrees for 30-35 minutes, 40-45 minutes for Bundt pan.

Per serving (excluding unknown items): 379.2 Calories; 20.5g Fat (47.1% calories from fat); 7.0g Protein; 44.9g Carbohydrate; 72 mg Cholesterol; 269 mg Sodium Exchanges: 1 ½ Grain (Starch); ½ Lean Meat; 1 ½ Fruit; 4 Fat; 1 ½ Other Carbohydrates.

# Spice Cake

Makes 8 servings. Prep Time: 45 minutes

**1 ¼ c whole wheat flour**
**1 c brown sugar**
**¼ c cornstarch**
**1 tsp cinnamon**
**1 tsp baking soda**
**½ tsp cloves**
**¼ tsp nutmeg**
**½ tsp salt**
**¼ tsp ginger**
**1 c water**
**3 tbsp oil**
**1 tbsp vinegar**

1. Combine dry ingredients with a fork in an 8x8 or round layer cake pan.
2. Make a well in middle. Pour in liquids and stir with a fork or whisk until smooth.
3. Bake at 350 degrees for about 30 minutes.
4. Cool on rack. Remove from pan after 10 minutes.
5. Makes one square or round layer cake.

*Mix-in-the-pan cakes save on work and cleanup!*

Good frosted or served with applesauce topping.

---

Per serving (excluding unknown items): 229.4 Calories; 5.5g Fat (20.9% calories from fat); 2.6g Protein; 44.5g Carbohydrate; 0mg Cholesterol; 304mg Sodium. Exchanges: 1 Grain(Starch); 1 Fat; 2 Other Carbohydrates.

# Stovetop Upside-Down Cake

Makes 10 servings. Prep Time: 45 minutes

**1 each Plain Vanilla Cake batter**
**2 tbsp oil**
**2 tbsp shortening**
**½ c brown sugar**
**1 c canned peaches**

1. Heat shortening and oil in a heavy 10-inch, straight-sided skillet until shortening melts.
2. Sprinkle brown sugar on top; arrange fruit on top of sugar.
3. Pour cake batter carefully over all and cook, covered, on medium-low heat for 25-30 minutes, until cake is set.
4. Immediately invert onto heatproof plate. Let skillet remain on a few minutes before removing, to allow topping to drip onto cake.
5. Serve while warm. Makes one 10-inch layer cake.

**Baked Cake**: Prepare as above. Use a 9-inch layer cake pan. Bake in a 350 degree oven for 30-40 minutes.

*This is a scaled-down, don't-have-an-oven version of a pineapple upside-down cake. Any available fruit, canned or fresh may be used.*

Per serving (excluding unknown items): 409.3 Calories; 11.5g Fat (29.3% calories from fat); 3.2g Protein; 59.3g Carbohydrate; 45mg Cholesterol; 383mg Sodium. Exchanges: ½ Fruit; 2 ½ Fat; 3 ½ Other Carbohydrates.

# Whole Wheat Coffee Cake

Makes 8 servings. Prep Time: 50 minutes

**5 tbsp shortening**
**2/3 c brown sugar**
**1 ½ tbsp soy flour**
**1 tsp vanilla**
**1 c water**
**1 ½ c whole wheat flour**
**1 tsp baking powder**
**¼ tsp salt**
**3 tbsp powdered milk**
**1/3 c brown sugar**
**1 ½ tbsp whole wheat flour**
**1 ½ tbsp shortening**
**½ tsp cinnamon**
**1/3 c rolled oats**

1. Cream shortening and 2/3 c brown sugar. Add vanilla and soy flour and stir well.
2. Stir together 1 ½ c flour, baking powder, salt and powdered milk in a separate bowl.
3. Add alternately with water to creamed mixture, beating thoroughly.
4. Pour into a greased and floured 9x9 pan.
5. Make topping by combining brown sugar, 1 ½ tbsp whole wheat flour, 1 ½ tbsp shortening, cinnamon and rolled oats.
6. Sprinkle on topping. Bake at 350 degrees for 25-30 minutes. Makes one 9x9 square pan.

Per serving (excluding unknown items): 311.0 Calories; 12.1g Fat (33.8% calories from fat); 5.0g Protein; 48.3g Carbohydrate; 3mg Cholesterol; 136mg Sodium. Exchanges: 1 ½ Grain(Starch); 2 ½ Fat; 2 Other Carbohydrates.

# Granny Hall's Fried Pies

Makes 1 serving. Prep Time: 2 hours

---

**sugar**
**dried fruit**
**plenty of shortening**
**biscuit dough or pastry**

1. Any dried fruit, such as peaches, apples or apricots may be used.
2. Soak dried fruit several hours or overnight. Stew the fruit in the soak water. As the fruit gets tender, add sugar to taste and a pinch of salt.
3. When fruit is thoroughly tender, mash amount to use in pies—about 1 heaping tbsp per pie (more if you want fat pies.) Have this cool before putting into pie crusts. It should be sweet and about the consistency of jam.
4. Roll out dough, not too thin, but as thin as you think will hold the fruit.
5. Cut pie crusts with small saucer or cereal bowl into circles. Re-roll edges to make more.
6. Place fruit in one half of the dough, spread out a bit, then fold over other half of the circle, moistening the inner edges. Use fork or fingers to press the curved edges together.
7. Heat shortening in skillet, medium hot. Using pancake turner, ease pie into hot fat, browning thoroughly on one side, then the other. You may stand the pies up on their curved edge to get the edges brown, or fry in deeper fat.
8. Carefully remove pies with pancake turner. Serve warm.

*This is an old Southern family recipe from my husband's side of the family. It came from a family cookbook entitled, "In Praise of Fried Pies."*

---

Per serving (excluding unknown items): 0.0 Calories; 0.0g Fat (0.0% calories from fat); 0.0g Protein; 0.0g Carbohydrate; 0.0mg Cholesterol; 0.0mg Sodium. Exchanges: Free.

# Granola Piecrust

Makes 8 servings. Prep Time: 20 minutes

**2 c granola, crushed**
**2 tbsp sugar**
**¼ c shortening, melted**

1. Combine crushed granola, sugar and melted shortening thoroughly.
2. Spread mixture in 9-inch pie pan and press firmly onto bottom and sides. This may be done by pressing another pie pan down into the one containing the crust.
3. Bake at 350 degrees for 6-8 minutes. Use as you would a graham cracker crust.

**Oat-Apple Piecrust**: Do not use above recipe. Instead, combine in a medium bowl 1 ½ c. grated apple, 1 c. rolled oats and ½ tsp salt. Mix and press into pie pan. Fill with heated, uncooked filling and bake as for other pies.

Per serving (excluding unknown items): 217.3 Calories; 14.7g Fat (58.3% calories from fat); 3.8g Protein; 20.0g Carbohydrate; 0mg Cholesterol; 3mg Sodium. Exchanges: 1 Grain(Starch); 3 Fat.

# Popcorn Piecrust

Makes 8 servings. Prep Time: 20 minutes

**2 ½ c popcorn, medium-ground**
**½ c shortening, melted**
**¼ c brown sugar**
**¼ tsp cinnamon**

1. Lightly grease a 9-inch pie pan.
2. Combine all ingredients thoroughly.
3. Spread evenly in pie pan. Press firmly onto bottom and sides of pan. This is easily done with a second pie pan pressed into the one containing the piecrust.
4. May be used uncooked or pre-baked at 350 degrees for 6-10 minutes, until lightly browned.

*This popcorn piecrust may be used like a graham cracker crust!*

Per serving (excluding unknown items): 156.5 Calories; 13.8g Fat (77.4% calories from fat); 0.3g Protein; 8.7g Carbohydrate; 0mg Cholesterol; 33mg Sodium. Exchanges: 3 Fat; ½ Other Carbohydrates.

# Standard Whole Wheat Piecrust

Makes 8 servings. Prep Time: 15 minutes

**6 ½ tbsp shortening**
**1 c whole wheat flour**
**½ tsp salt**
**2 tbsp cold water**

1. Stir together flour and salt. Cut in shortening using pastry blender or two knives until particles are the size of small peas.
2. Sprinkle water, one tbsp at a time, and toss with a fork until all flour is moistened (an extra tbsp or so of water may be needed.)
3. Gather pastry into a ball and then flatten on a lightly floured cloth-covered board or kitchen counter.
4. Roll 2 inches larger than inverted pie plate. Fold in half to move onto pie plate. Then unfold and press firmly against bottom and side of pie plate.
5. Makes one bottom crust for an 8-9-inch pie.

**Waxed Paper Alternative**: Pastry may be rolled out between sheets of waxed paper to minimize cleanup, or to make transfer to pie plate easier. Pre-wipe counter with a damp cloth to prevent waxed paper from slipping. Tear off two square sheets of waxed paper. Place one on damp counter. Place pastry ball on waxed paper and flatten. Cover with second piece of waxed paper. Roll out to desired size. Remove top sheet of waxed paper. Transfer to pie plate by flipping over pastry so bottom waxed paper is on top. Remove bottom paper and proceed as above.

**Pre-Baked Piecrust**: If a pre-baked shell is required, prick pastry thoroughly with fork, then bake at 475 degrees for 8-10 minutes.

*Soft white wheat, all-purpose flour, or half-and-half mixture with whole wheat, will produce a more tender and lighter-colored piecrust.*

Use in any fruit or custard pie recipes.

---

Per serving (excluding unknown items): 142.9 Calories; 10.7g Fat (65.0% calories from fat); 2.1g Protein; 10.9g Carbohydrate; 0mg Cholesterol; 134mg Sodium. Exchanges: ½ Grain(Starch); 2 Fat.

# Whole Wheat Oil Piecrust

Makes 8 servings. Prep Time: 15 minutes

**1 c whole wheat flour**
**½ tsp salt**
**$^1/_3$ c oil**
**2 tbsp cold water**

1. Mix flour, oil and salt with a fork until particles are the size of small peas.
2. Add water and mix as for standard piecrust.
3. Form using waxed paper instructions.
4. If a pre-baked shell is required, prick thoroughly and bake at 475 degrees for 12-15 minutes.
5. Pre-baked or not, cover crust edges with strips of aluminum foil to prevent burning during baking.
6. Makes one bottom crust for an 8-9-inch pie.

*Soft white wheat or all-purpose flour, or a half-and-half mixture with whole wheat, will produce a more tender and lighter-colored piecrust.*

Use as any standard piecrust, for fruit or custard fillings

Per serving (excluding unknown items): 131.1 Calories; 9.4g Fat (61.9% calories from fat); 2.1g Protein; 10.9g Carbohydrate; 0mg Cholesterol; 134mg Sodium.
Exchanges: ½ Grain(Starch); 2 Fat.

# Lentil Fudge Pie

Makes 8 servings. Prep Time: 1 hour and 10 minutes

**4 tbsp cocoa**
**4 tbsp shortening, melted**
**¼ c sugar**
**¾ c light corn syrup (see note)**
**3 med eggs, reconstituted**
**1 tsp vanilla**
**1 ¾ c cooked lentils**
**1 whole pie crust (9 inch), unbaked**

1. Combine cocoa and melted shortening. Mix in sugar, corn syrup, eggs and vanilla, blending for 2 minutes.
2. Gently stir in cooked lentils.
3. Pour into unbaked pie shell. Bake at 375 degrees for 40-50 minutes, until knife inserted in center comes out clean.
4. Cool completely before serving.

NOTE: Make corn syrup replacement by heating together ½ c water and 1 ½ c sugar until sugar is dissolved. Measure out required amount for recipe. Allow to cool before adding.

*This is really fudge-y! The lentils almost look like mini-chocolate chips and they absorb the fudge flavor.*

---

Per serving (excluding unknown items): 353.8 Calories; 14.9g Fat (36.4% calories from fat); 8.1g Protein; 50.5g Carbohydrate; 81mg Cholesterol; 192mg Sodium. Exchanges: 1 ½ Grain(Starch); ½ Lean Meat; ½ Fruit; 3 Fat; 2 Other Carbohydrates.

# Split Pea Pie

Makes 16 servings. Prep Time: 1 hour and 45 minutes

**1 c split peas, yellow**
**2 ½ c water**
**1 c sugar**
**½ tsp ginger**
**½ tsp salt**
**½ tsp cinnamon**
**¼ tsp nutmeg**
**¼ tsp cloves**
**1 tsp vanilla**
**3 med eggs, reconstituted**
**1 ⅓ c evaporate skim milk (see note)**
**2 whole pie crusts, 8-inch**

1. Simmer peas with water for 40 minutes in a covered pan. Do not drain.
2. Puree cooked peas and water; allow to cool.
3. Add sugar, spices, vanilla and eggs. Thoroughly blend with whisk. Add milk and blend another 30 seconds.
4. Heat oven to 375 degrees. Place unbaked pie shells in oven, then carefully pour in filling.
5. Bake 40 minutes, until knife inserted in center comes out clean.
6. Serve warm or cool. Makes 2 8-inch pies.

NOTE: Make evaporated milk from powdered milk by combining 1 ¼ c water and ½ c powdered milk.

*Another surprising and delicious use for legumes!*

---

Per serving (excluding unknown items): 122.3 Calories; 1.1g Fat (8.3% calories from fat); 5.8g Protein; 22.6g Carbohydrate; 41mg Cholesterol; 106mg Sodium. Exchanges: ½ Grain(Starch); ½ Lean Meat; 1 Fruit; 1 Other Carbohydrates.

# Thanksgiving Pie

Makes 8 servings. Prep Time: 1 hour and 30 minutes

**2 c pinto bean puree**
**3 med eggs, reconstituted**
**1 ½ c evaporate skim milk (see note)**
**1 c sugar**
**½ tsp salt**
**1 tsp cinnamon**
**1 tsp ginger**
**¼ tsp cloves**
**¼ tsp nutmeg**
**1 whole pie crust, 9-10 inch**

1. Preheat oven to 375 degrees.
2. In a large bowl, beat eggs with a whisk. Add bean puree, milk and sugar. Stir with whisk until smooth.
3. Add salt and spices. Stir until blended.
4. Place pie shell (unbaked) in oven, then carefully pour in filling.
5. Bake about 60 minutes, or until knife inserted in center comes out clean.
6. Serve warm or cool. Makes one 9-10-inch pie.

NOTE: Make evaporated milk from powdered milk by combining 1 ⅓ c water with ½ c plus 1 tbsp powdered milk.

*See if your holiday guests can tell this from the real thing!*

Per serving (excluding unknown items): 383.2 Calories; 8.3g Fat (19.1% calories from fat); 10.5g Protein; 68.3g Carbohydrate; 83mg Cholesterol; 418mg Sodium. Exchanges: 2 ½ Grain(Starch); ½ Lean Meat; ½ Non-Fat Milk; 1 ½ Fruit; 1 ½ Fat; 1 ½ Other Carbohydrates.

# Fresh Fruit Pie

Makes 8 servings. Prep Time: 1 hour and 30 minutes

**2 ea pie crust (9 inch)**
**¾ c sugar**
**¼ c all-purpose flour**
**½ tsp nutmeg**
**½ tsp cinnamon**
**dash salt**
**6 med apples, peeled, thinly sliced**

1. Prepare pie crusts. Heat oven to 425 degrees.
2. Mix together dry ingredients. Stir in apples. Turn into bottom pie crust.
3. Cover with top crust and cut several small slits in it to allow steam to escape. Crimp edges together.
4. Cover edges with aluminum foil to prevent crust from burning.
5. Bake 40-50 minutes, until crust is brown and juice bubbles through slits.

**French Apple Pie**: Omit top crust. Instead, use 1 recipe of fruit crisp topping. Bake as above.

**Blueberry Pie**: Use ½ c sugar, 1/3 c flour and 4 c blueberries. Omit nutmeg and salt.

**Plum Pie**: Same as blueberry, but use 4 c sliced plums.

**Strawberry Pie**: Same as blueberry, but use 4 c sliced strawberries.

**Peach Or Apricot Pie**: Use 1 c sugar, ¼ c flour and 5 c sliced peaches. Omit nutmeg and salt.

**Rhubarb Pie**: Use 1 ½ c sugar, 1/3 c flour and 4 c ½-inch sliced rhubarb. Omit nutmeg, cinnamon and salt.

**Cherry Pie**: Same as rhubarb, but use 4 c canned or fresh pie cherries. Add ¼ tsp almond extract.

*Cornstarch may be used instead of all-purpose flour. Use half the amount.*

---

Per serving (excluding unknown items): 343.0 Calories; 12.6g Fat (32.2% calories from fat); 3.2g Protein; 56.3g Carbohydrate; 0mg Cholesterol; 292mg Sodium. Exchanges: 1 ½ Grain(Starch); 2 Fruit; 2 ½ Fat; 1 ½ Other Carbohydrates.

# Strawberry Glace Pie

Makes 8 servings. Prep Time: 45 minutes

**1 ea pie crust (9 inch), baked**
**6 c strawberries, fresh**
**1 c sugar**
**3 tbsp cornstarch**
**½ c water**

1. Mash enough strawberries to make one cup.
2. Mix together sugar and cornstarch in a 2-quart saucepan. Stir in water and strawberries gradually.
3. Cook over medium heat, stirring constantly, until mixture thickens and boils. Boil and stir one minute. Cool.
4. Fill baked pie shell with remaining strawberries. Pour cooked berry mixture over top so it runs through the berries. Cool until set.

**Peach Glace Pie**: Same as above, but peaches may need treatment to prevent darkening.

**Raspberry Glace Pie**: Same as above, substitute 6 cups raspberries.

Per serving (excluding unknown items): 241.9 Calories; 6.5g Fat (23.4% calories from fat); 2.0g Protein; 45.6g Carbohydrate; 0mg Cholesterol; 148mg Sodium. Exchanges: 1 Grain(Starch); 2 Fruit; 1 Fat; 1 ½ Other Carbohydrates.

# APPENDIX

- Four Week Meal Plan
- Meal Planning Guidelines
- Four Week Grocery List

# FOUR WEEK MEAL PLAN

Four weeks of sample meal plans have been provided. Only stovetop recipes have been used, so that it may apply to all situations. The recipes provide 2000-2500 calories per day and 25-35% of calories from fat.

The shopping list is based on using the meal plans for a family of four. If your family is larger or smaller than this, you will need to multiply or divide accordingly. Amounts for many of the foods have been added together and rounded to amounts and units in which they are usually purchased. Other very small amounts have been left in their original form, so that you can easily tell if you have sufficient on hand already. Ingredients for prepared foods have been listed as individual ingredients. For example, bread has been broken down into its ingredients: yeast, salt, oil, sugar, flour, etc.

All entries that begin with a capital letter are recipes from this book and may be looked up in the recipe index. Some have had to be abbreviated due to space limitations, but they should still be easily distinguished. Those entries that begin with a lower case letter and an amount are individual food items for which there is no recipe (such as canned peaches, sugar or dill pickles.) Serving sizes for these are listed in the meal plan. Serving sizes for the recipes are listed in the actual recipes.

Since some fresh fruits and vegetables are used in the meal plan, canned or dried alternatives would need to be substituted if fresh is not available.

To save on fuel and meal preparation time, some of the meals use pre-cooked foods from a previous meal. Beans, rice or bread, prepared for one meal may be cooked in quantity and used for a subsequent meal. Extra should be prepared at the first meal to provide for both. Those first meals that need extra food prepared are starred once, and the meals that use those foods are starred twice. Be sure to look for the stars when planning your cooking for the day.

## Week One

| Day | Breakfast | Lunch | Dinner |
|---|---|---|---|
| 1 | Pancakes<br>Maple-flavored Syrup<br>Milk | Cowpuncher Stew*<br>Panned Bread*<br>1 apple<br>Milk | Bean Stroganoff**<br>½ c. green beans<br>Granny's Fried Pies**<br>Milk |
| 2 | Popcornmeal<br>1 tbsp sugar<br>2 tbsp raisins<br>Milk | Spiced Cheese*<br>Tortillas<br>½ c. canned peaches<br>Milk | Fried Cheese**<br>Garlic Rice and Pasta<br>Sauteed Greens<br>Raspberry Dumplings<br>Milk |
| 3 | Skillet Toast*<br>2 tbsp jelly<br>Milk | Onion-lentil Soup*<br>Dumplings<br>2 plums<br>Milk | Rice Patties<br>Panned Bread**<br>Lentil Salad**<br>No-bake Choco-PB bars<br>Milk |
| 4 | Morning rice<br>1 tbsp sugar<br>Milk | Tuna Spread Sandwiches<br>Panned Bread*<br>1 dill pickle<br>½ c. applesauce<br>Milk | Speedy Spaghetti<br>Panned Bread**<br>1 c. salad greens<br>French Dressing<br>Fried AppleRings<br>Milk |
| 5 | Potato Pancakes<br>½ c. applesauce<br>Milk | Basic Bean Burrito<br>Fresh Salsa<br>Popcorn Variations<br>½ c. canned peaches<br>Milk | Beef and Barley Soup<br>Basic Popcorn<br>Upside Down Cake<br>Milk |
| 6 | Oatmeal<br>1 tbsp sugar<br>2 tbsp raisins<br>Milk | Easy Lentil Chili<br>Corn Bread<br>½ c. canned plums<br>Milk | Stir Fry Dinner<br>Stovetop Rice<br>Stewed Fruit<br>Milk |
| 7 | Biscuits and Gravy<br>Milk | PB and What? Sandwich<br>Carrot Casserole<br>1 apple<br>Milk | Grand Prize Zucch. Skillet<br>Potato Planks<br>Molasses Crunch Popcorn<br>Milk |

## Week Two

| Day | Breakfast | Lunch | Dinner |
|---|---|---|---|
| 1 | Cream of Rice<br>1 tbsp sugar<br>2 tbsp raisins<br>Milk | Black Bean and Rice Salad<br>Panned Bread*<br>½ c. canned pears<br>Milk | Marco Polo Spaghetti<br>Panned Bread Sticks**<br>½ c. cooked spinach<br>Cake Doughnuts<br>Milk |
| 2 | Fried Mush<br>Maple-flavored Syrup<br>Milk | Tuna Patties<br>Panned Bread Buns**<br>1 dill pickle<br>½ c. canned cherries<br>Milk | Spanish Rice<br>½ c. canned green beans<br>Raspberry Dumplings<br>Milk |
| 3 | Basic Breakfast Popcorn<br>1 tbsp sugar<br>½ c. canned pears<br>Milk | Lentil Stew<br>Dumplings<br>½ c. applesauce<br>Milk | Middle East Sandwiches<br>Tortillas<br>Fresh Fruit Compote<br>Milk |
| 4 | English Muffins*<br>2 tbsp jam<br>Milk | Mini Pizzas**<br>Italian Potatoes<br>½ c. canned peaches<br>Milk | Classic Split Pea Soup<br>Steamed Dinner Loaf<br>Cooked Dried Fruit<br>Milk |
| 5 | Lumpy Dick<br>1 tbsp sugar<br>1 tsp cinnamon<br>Milk | Potato Soup<br>Croutons<br>Autumn Carrots<br>½ c. canned plums<br>Milk | Spanish Wheat<br>Lentil Salad<br>Chewy Goober Bars<br>Milk |
| 6 | Squash Griddle Cakes<br>2 tbsp honey<br>Milk | Chicken Chili*<br>Tortilla Chips*<br>½ c. applesauce<br>Milk | Hot Mexican Salad**<br>Tortilla Chips**<br>Vanilla Pudding<br>Milk |
| 7 | Hashed Browns<br>½ c. applesauce<br>Milk | South-of-the-Border San**<br>Popcorn Variations<br>1 c. salad greens<br>Salad Dressing<br>Milk | Macaroni Skillet<br>½ c. canned tomatoes<br>Steamed Chocolate Cake<br>Milk |

## Week Three

| Day | Breakfast | Lunch | Dinner |
|---|---|---|---|
| 1 | Muesli<br>Milk | Corn Chowder<br>Dumplings<br>½ c. canned cherries<br>Milk | Sloppy Joes<br>Panned Bread Buns<br>½ c. canned green beans<br>Honey-Popcorn Balls<br>Milk |
| 2 | Irish Soda Scones<br>2 tbsp honey<br>Milk | Italian Pea Soup<br>Popcorn Variations<br>½ c. applesauce<br>Milk | Browned Rice<br>Refried Beans<br>½ c. canned tomatoes<br>No-bake PB Cookies<br>Milk |
| 3 | Wheat Berry Cereal*<br>1 tbsp sugar<br>2 tbsp raisins<br>Milk | Quick Wholewheat Chili**<br>Polenta<br>½ c. peaches<br>Milk | Lentil Burgers<br>½ c. mashed potatoes<br>Apple Fritters<br>Milk |
| 4 | Skillet Fries<br>½ c. canned pears<br>Milk | Pasta Salad<br>Hush Puppies<br>½ c. canned plums<br>Milk | Spanish Bean Pot<br>Hoe Cakes<br>Steamed Carrot Pudding<br>Milk |
| 5 | Granola<br>Milk | East Indian Cheese & Veg*<br>Tortillas<br>½ c. applesauce<br>Milk | Under Wraps**<br>(use cheese)<br>½ c. canned corn<br>Cake Doughnuts<br>Milk |
| 6 | Boston Brown Bread<br>½ c. applesauce<br>Milk | Beefed Up Sandwiches<br>(uses Panned Bread)*<br>Popcorn Variations<br>½ c. peaches<br>Milk | Speedy Spaghetti<br>Panned Bread**<br>Cabbage Wedges<br>Stewed Apples<br>Milk |
| 7 | Navajo Fry Bread*<br>1 tbsp sugar<br>1 tsp cinnamon<br>Milk | Bean Bonanza*<br>Steamed Nut or Oat Bread<br>½ c. canned plums<br>Milk | Tuscany Soup**<br>Croutons**<br>Steamed Pumpkin Bread<br>Milk |

## Week Four

| Day | Breakfast | Lunch | Dinner |
|---|---|---|---|
| 1 | Cracked Wheat Cereal*<br>1 tbsp sugar<br>2 tbsp raisins<br>Milk | Barbecue Sandwiches<br>Panned Bread<br>½ c. canned pears<br>Milk | Wheat Verde**<br>Refried Beans*<br>Fresh Salsa<br>No-bake Choc. PB Bars<br>Milk |
| 2 | Oatmeal Pancakes<br>Maple-flavored Syrup<br>Milk | Stovetop Rice*<br>Refried Beans**<br>Fresh Salsa<br>½ c. applesauce<br>Milk | Lentil Burgers<br>Mashed Potatoes*<br>½ c. pickled beets<br>Upside-Down Cake<br>Milk |
| 3 | Honey-Frosted Popcorn<br>½ c. canned pears<br>Milk | Lentils and Greens Soup<br>Tatertots**<br>½ c. canned plums<br>Milk | Pita and Falafel Filling<br>(use panned bread or tortillas instead of pita)<br>Coleslaw<br>Chocolate Pudding<br>Milk |
| 4 | Skillet toast<br>2 tbsp jelly<br>Milk | PB and What? Sandwiches<br>Kraut and Apple Salad<br>Popcorn Variations<br>Milk | Red Beans and Rice*<br>½ c. canned green beans<br>Stewed Fruit<br>Milk |
| 5 | Morning Rice<br>½ c. applesauce<br>Milk | Mediterran. Barley Salad<br>Soybean Patties**<br>(use leftover red beans)<br>½ c. canned peaches<br>Milk | Homemade Noodles<br>½ c. canned tomatoes<br>Basic Cooked Beans<br>Molasses Crunch Popcorn<br>Milk |
| 6 | Lumpy Dick<br>1 tbsp sugar<br>1 tsp cinnamon<br>Milk | Four-Alarm Chili*<br>Tortillas*<br>½ c. canned pears<br>Milk | East Indian Cheese,Spinach<br>Tortillas**<br>Raspberry Dumplings<br>Milk |
| 7 | Cake Doughnuts<br>½ c. applesauce<br>Milk | Vegetarian BLT Sandwich<br>(use Panned Bread)**<br>Chili Bean Salad<br>(use leftover beans)<br>½ c. canned cherries<br>Milk | Almost Chicken Soup<br>Dumplings<br>Fried Apple Rings<br>Milk |

## Meal Planning Guidelines

To plan your own menus that are nutritionally well-balanced, there are a few guidelines to keep in mind. The accompanying figure shows all foods divided into five basic groups based on what nutrients they provide. Starches (breads, cereals, etc.) provide energy and B vitamins. Vegetables and fruits provide health-protecting vitamins such as folacin and vitamins A and C. The milk group provides protein and calcium. The meat, beans, etc. group also provides protein. The fats, oils and sweets group does not really count as a food group at all, but is a reminder that they should be used sparingly, after other health-promoting foods have been eaten in sufficient quantities. The tiny circles and triangles in each of the food groups is also a reminder of the small amounts of hidden fats and sugars in all foods. Time spent studying this figure before beginning meal planning would be well spent.

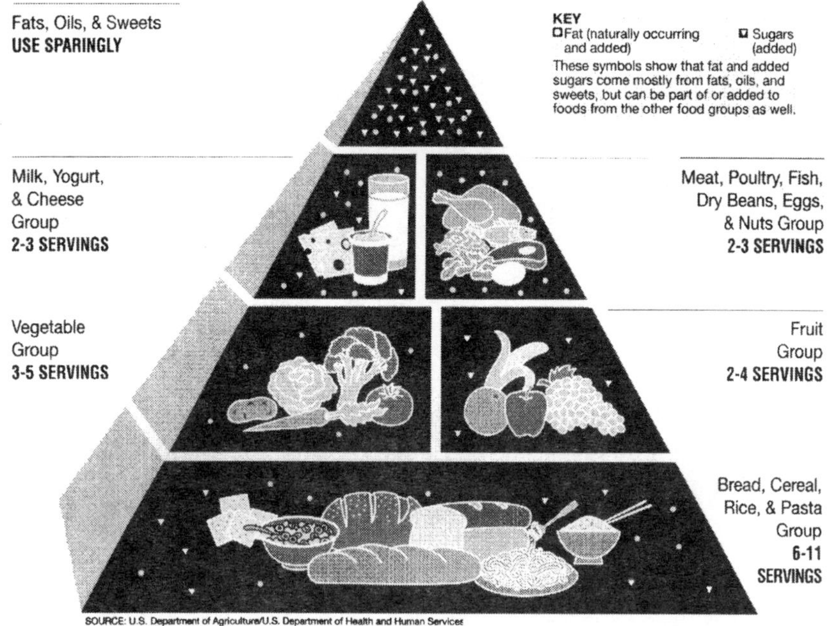


The bottom of the Food Guide Pyramid shows that starches should be the foundation of an adequate diet. As you go up the pyramid, the blocks become smaller, and so do the number of servings per day needed from that food group. Each food group has a range of suggested servings per day. The greater number of servings would be for those individuals who require more calories (teens, tall people, pregnant or lactating women, those with high activity levels). Serving size is generally ½ cup or 1 slice for starches, ½ cup for fruits and vegetables, 1 cup for milk and ½ cup for beans (2 oz. for meat or fish if these are available).

## Basic Meal Pattern

In writing the four-week meal plan, the following pattern was used, and it may also be used to plan your own meals for your family. This is only a basic pattern, and the number of servings of each food group may be adjusted for each individual's needs.

Breakfast: 1 starch, 1 milk, 1 fruit
Lunch: 2 starch, 1 bean, 1 vegetable, 1 fruit, 1 milk
Dinner: 2 starch, 1 bean, 2 vegetable, 1 fruit, 1 milk, 1 sweet

Since most of the main dishes include foods from several food groups, one dish may count for several servings. For example, Almost Chicken Soup with Rice provides 1 cup cooked rice (2 starch servings), ¼ cup TVP (1 meat/bean serving) and ½ cup of several vegetables (1 vegetable serving.)

## Using Diabetic Exchanges

The exchanges shown in the nutrition information at the end of each recipe, although actually provided for diabetic meal planning, can help with family meal planning as well. There is only one main difference between the diabetic exchanges and the food guide pyramid that needs to accounted for when using them. That difference is the meat serving size. Almost Chicken Soup with Rice is shown to have 2 ½ meat exchanges in its nutrition information. The reason for this difference is that diabetics count each ounce as an exchange, whereas the food guide pyramid counts 2-3 ounces as a serving. The "other carbohydrate" exchange listed in foods that contain a significant amount of sugar (see any dessert recipe for an example) refers to sugar content of the dish and is to help diabetics to control their sugar intake.

## Vitamins A and C

To be sure that intakes of vitamins A and C are adequate, one fruit/vegetable serving per day should be a good source of vitamin C, and one serving every other day sould be a good source of vitamin A. Vitamin C foods include potatoes, cabbage, tomatoes and strawberries. Vitamin A foods include carrots, yellow winter squash, pumpkin, greens, spinach and broccoli. Using all the above guidelines, you may feel confident that you are doing your best to provide your family with a healthy diet off the grid!

# FOUR WEEK GROCERY LIST

| | |
|---|---|
| 4 lb | all-purpose flour |
| 2 tsp. | allspice |
| 1 lb | any dried fruit |
| 26 whole | apples |
| 5 quarts | applesauce |
| 2 lb | baking powder |
| 8 oz | baking soda |
| 2 3/4 tbsp. | basil |
| 6 each | bay leaves |
| 4 oz | beef bouillon granules |
| 1 lb | black beans |
| 1 lb | broccoli |
| 1 lb | brown sugar |
| 2 heads | cabbage |
| 7 quarts | canned tomatoes |
| 1 1/3 tbsp. | caraway seed |
| 18 large | carrots |
| 16 oz | catsup |
| 3/8 tsp. | cayenne |
| 2 quarts | cherries in syrup |
| 5 oz | chickn bouillon granules |
| 1 oz | chili powder |
| 2/3 tbsp. | chives |
| 6 oz | chocolate chips |
| 1 1/2 lb | chopped nuts |
| 4 tbsp. | cilantro |
| 2 sticks | cinnamon |
| 2 7/8 tbsp. | cinnamon |
| 1/8 tsp | cloves |
| 16 whole | cloves |
| 1 lb | cocoa |
| 21 cups | cooked rice |
| 1 1/2 tsp. | coriander |
| 1 quart | corn flour |
| 4 cans | corn, canned |
| 3 lb | cornmeal |
| 1 lb | cornstarch |
| 1 lb | cracked wheat |
| 2 3/8 tbsp. | cumin |
| 8 whole | dill pickles |
| 2 tsp | dill weed |
| 3/4 tsp. | dry mustard |
| 1/2 lb | eggs, powdered |
| 1 lb | elbow macaroni |
| 1/4 tsp. | fennel seed |
| 2 quarts | fresh raspberries |
| 2 oz | garlic powder |
| 2 lb | garbanzo beans |
| 1/8 tsp | ginger |
| 3 quarts | green beans, canned |
| 13 large | green bell pepper |
| 12 med | green chiles |
| 3 cups | green onion |
| 2 pints | honey |
| 5 medium | jalapenos |
| 1 cup | jam |
| 2 pints | jelly |
| 5 lb | kidney beans |
| 7 lb | lentils |
| 3 tbsp. | maple flavoring |
| 2 tsp. | marjoram |
| 1/3 cup | mashed pumpkin |
| 1 1/2 cups | mayo-type dressing |
| 1/2 tsp | mint |
| 6 quarts | mixed salad greens |
| 1 quart | mixed vegetables |
| 12 oz | molasses |
| 1 lb | navy beans |
| 7/8 tsp. | nutmeg |
| 3 quarts | oil |
| 8 oz | olive oil |
| 15 lb | onion |
| 2 5/8 tbsp. | onion flakes |
| 1 1/8 tbsp. | oregano |

| | |
|---|---|
| 5/8 tsp. | paprika |
| 2 cups | parsley |
| 1/3 lb | pasta, any large shape |
| 5 quarts | peach halves in syrup |
| 2 lb | peanut butter |
| 4 quarts | pear halves in syrup |
| 1 lb | pearl barley |
| 1 pint | peas |
| 2 tbsp. | pepper |
| 1 pint | pickled beets |
| 5 lb | pinto beans |
| 2 pounds | plums |
| 4 quarts | plums in syrup |
| 4 lb | popcorn |
| 1 lb | potato flakes |
| 15 lb | potatoes |
| 25 lb | powdered milk |
| 1/4 cup | powdered sugar |
| 3 lb | raisins |
| 5 large | red bell pepper |
| 1/8 tsp | red pepper flakes |
| 3 lbs | rice |
| 3 lb | rolled oats |
| 1/3 tsp | sage |
| 1 lb | salt |
| 1 qt | sauerkraut |
| 2 tbsp. | sesame seeds |
| 6 lb | shortening |
| 3 whole | cucumbers |
| 1 lb | soy bacon bits |
| 1 lb | soy flour |
| 8 oz | soy sauce |
| 1 lb | soybeans |
| 4 lb | spaghetti |
| 3 lb | spinach |
| 1 pint | spinach, canned |
| 1 lb | split peas |
| 15 lbs | sugar |
| 2 tbsp. | sweet pickle relish |
| 3 lb | Swiss chard |
| 5 1/3 dashes | Tabasco sauce |
| 2 lb | textured soy protein |
| 1 5/8 tsp. | thyme |
| 18 med | tomatoes |
| 1 1/2 cups | tomato juice |
| 1 tbsp. | tomato paste |
| 2 pints | tomato sauce |
| 6 cans | tuna in water |
| 1/2 tsp. | turmeric |
| 1 each | turnip |
| 1 3/4 tbsp. | vanilla |
| 2 quart | vinegar |
| 1 pt | wax beans, canned |
| 3 lb | whole wheat berries |
| 25 lb | whole wheat flour |
| 1 lb | winter squash |
| 4 oz | Worcestershire sauce |
| 1 lb | yeast |
| 3 med | zucchini |

# ALPHABETICAL RECIPE LIST

# INDEX